Marshall Fishwick

Popular Culture
Cavespace to Cyberspace

Pre-publication REVIEWS, COMMENTARIES, EVALUATIONS . . .

"Marshall Fishwick takes us on a twisted, leaping romp through time, creating a vivacious manifesto for popular culture. Along the way, he asks questions, serious questions in jaunty prose, and they must be answered if we are to recapture our sanity."

Henry Glassie, PhD
Professor, Folklore Institute, Indiana University, Bloomington, IN

"This book enables a well-traveled, well-read author to bring together a number of essays on places, politics, and events that have all had or are still having an effect on popular culture.

You will visit Bangladesh and risk your life in the traffic, but how will the population deal with the electronic revolution where 90 percent of the people eke out a rural existence? Sports are compared to religion. From cave paintings to Coca Cola bottles, from current American foreign policy to the history of McDonald's arches, the choice of coverage is wide. As a true academician, Marshall Fishwick ends each essay with a bibliography for those who want to follow up the background of this book and his excellent research."

John Griffiths, MA
Associate of the Royal Institute of British Architects; Chartered Architect and Freelance Writer

More pre-publication
REVIEWS, COMMENTARIES, EVALUATIONS . . .

"Marshall Fishwick writes that 'The more we study popular culture the more baffling and unpredictable it becomes,' and 'Revolution is popular culture going top-speed—"voice of the people" shouting.' This book will inform and surprise you. Popular culture seems all encompassing, but it is not mass culture. It is ancient, medieval, modern and contemporary, American and worldwide—and it is not all good: see for instance Fishwick's description of his experience in Bangladesh. His observations are keen; his insights, deep. Here you will find social and intellectual history along with current political and cultural analysis. This is a wide-ranging effort. Among other personalities and subjects are Plato, Cicero, F.D.R., Henry Ford, Walt Disney, Edison, and Michael Jordan, plus philosophy, religion, and sports. And if you wish to examine them more carefully, endnotes conveniently placed at the close of each chapter include descriptions of relevant reading. Fishwick has done an extraordinary job of synthesis that should appeal to student, scholar, and general reader."

Marvin Wachman, PhD
Chancellor,
Temple University,
Philadelphia, PA

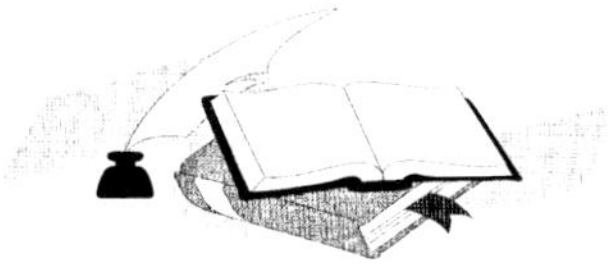

"In *Popular Culture: From Cavespace to Cyberspace,* Marshall Fishwick sets out 'neither to praise nor to bury popular culture—instead to try to understand its history, function and role.' Beginning with the great Greek philosophers such as Thales, Socrates, Plato, and Aristotle, Fishwick moves through age after age to post-modern times, using events and ideas to depict the common culture that has shaped the lives of people in every generation.

The ideas and insights of this book are shaped by a long career spent analyzing the common culture of the people, and communicating those insights to many generations of students. Fishwick brings impressive learning and stimulating insights as he sets himself the task of showing how popular culture has functioned through the ages in human society."

Harry B. Adams, DD
Professor of Pastoral Theology,
Yale Divinity School

NOTES FOR PROFESSIONAL LIBRARIANS AND LIBRARY USERS

This is an original book title published by The Haworth Press, Inc. Unless otherwise noted in specific chapters with attribution, materials in this book have not been previously published elsewhere in any format or language.

CONSERVATION AND PRESERVATION NOTES

All books published by The Haworth Press, Inc. and its imprints are printed on certified pH neutral, acid free book grade paper. This paper meets the minimum requirements of American National Standard for Information Sciences–Permanence of Paper for Printed Material, ANSI Z39.48-1984.

Popular Culture
Cavespace to Cyberspace

HAWORTH Popular Culture

Frank W. Hoffmann, PhD and B. Lee Cooper, PhD
Senior Editors

New, Recent, and Forthcoming Titles:

Arts & Entertainment Fads by Frank W. Hoffmann and William G. Bailey

Sports & Recreation Fads by Frank W. Hoffmann and William G. Bailey

Mind & Society Fads by Frank W. Hoffmann and William G. Bailey

Fashion & Merchandising Fads by Frank W. Hoffmann and William G. Bailey

Chocolate Fads, Folklore, and Fantasies: 1000+ Chunks of Chocolate Information by Linda K. Fuller

The Popular Song Reader: A Sampler of Well-Known Twentieth Century Songs by William Studwell

Great Awakenings: Popular Religion and Popular Culture by Marshall W. Fishwick

The Christmas Carol Reader by William Studwell

Media-Mediated Relationships: Straight and Gay, Mainstream and Alternative Perspectives by Linda K. Fuller

The National and Religious Song Reader: Patriotic, Traditional, and Sacred Songs from Around the World by William E. Studwell

Rock Music in American Popular Culture: Rock 'n' Roll Resources by B. Lee Cooper and Wayne S. Haney

Rock Music in American Popular Culture II: More Rock 'n' Roll Resources by B. Lee Cooper and Wayne S. Haney

The Americana Song Reader by William E. Studwell

Images of Elvis Presley in American Culture, 1977-1997: The Mystery Terrain by George Plasketes

Popular Culture: Cavespace to Cyberspace by Marshall W. Fishwick

Popular Culture
Cavespace to Cyberspace

Marshall W. Fishwick

The Haworth Press
New York • London • Oxford

The Haworth Press, Inc., 10 Alice Street, Binghamton, NY 13904-1580

Cover design by Monica L. Seifert.

Library of Congress Cataloging-in-Publication Data

Fishwick, Marshall William.
Popular culture : cavespace to cyberspace / Marshall W. Fishwick.
p. cm.
Includes bibliographical references and index.
ISBN 0-7890-0643-X (alk. paper).
1. Popular culture. 2. Popular culture—History. 3. Space and time. 4. Social evolution. 5. Cyberspace. I. Title.
CB430.F574 1999
306—dc21 98-42781
CIP

For Ann—and all the students
who have really been my teachers.

ABOUT THE AUTHOR

Marshall Fishwick, PhD, Professor of Interdisciplinary Studies and Director of American Studies and Popular Culture at Virginia Tech in Blacksburg, Virginia, holds numerous honorary degrees and teaching awards. In 1998, Dr. Fishwick was honored by the American Culture Association (ACA) with its Award for Lifetime Achievement. Dr. Fishwick is co-founder of the Popular Culture Association (PCA). Having served as Fulbright Distinguished Professor in Denmark, Italy, Germany, Korea, and India, he was instrumental in establishing the American Studies Research Center in Hyderabad, India, which now houses the largest collection of American books in Asia. He is the author of *The Hero, American Style; American Heroes: Myth and Reality; American Studies in Transition; Faust Revisited; Icons of Popular Culture; Parameters of Popular Culture; Great Awakenings: Popular Religion and Popular Culture; Jane Addams; Clara Barton; Preview 2001; Go and Catch a Falling Star;* and *An American Mosaic: Rethinking American Culture History.* His latest book is *Globalism—Dead or Alive?* In addition, Dr. Fishwick serves as advisory editor of the *Journal of Popular Culture* and the *Journal of American Culture.* He helped found a new journal, *International Popular Culture.*

CONTENTS

Foreword

READING THE WINDS OF POPULAR CULTURE

Logically, a comprehensive history of the development of cultures should begin in the cave, because that is where human society began as popular culture and where some of the earliest records have been kept, waiting for people to come in with the expertise to read them.

Culture has always been consistent at the cave level because it consists of the everyday necessities of life. Through the millennia, it has changed as developments responded to differing physical aspects of life and people's reactions to them. As in architecture and other physical phenomena of life, form follows culture but it also leads it, creating a vacuum that sucks culture into new channels and over new horizons. But at the cave level—that is, the unaffected culture of most of the people most of the time—the culture has always been everyday, that is, popular in the sense of vox populi.

It takes a sensitive cultural meteorologist to chart the winds of cultural constancy and the shifts in direction and volume. Through the years popular culture has had many analysts, some qualified, many amateurish, who have posed as authorities on the world in which they live. It has always been easy for observers to hold up a wet finger and feel the breeze, or check a sensitive body joint, and reach conclusions about cultural weather in general. But such analyses, generally using tools that have been successful or a failure in other disciplines, generally have been superficial. They also have usually coddled a condescension based on inadequate information and self-interest bias as to how culture could be interpreted. Such incorrect analyses can bring dire results, for in interpreting the forces of culture, as with those of nature, a little knowledge is a dangerous thing.

The danger in our day comes from the fact that cultural analysts do not recognize or will not admit that culture is increasingly democratic, propelled by the forces that drive the growth of democracy, and must be read in that vocabulary. Today's popular culture consists of the many aspects of our individual and collective lives which we inherit from our forebears, use and modify as we will, and pass on to our descendants. The dynamics of this culture can perhaps best be illustrated in the graph below:

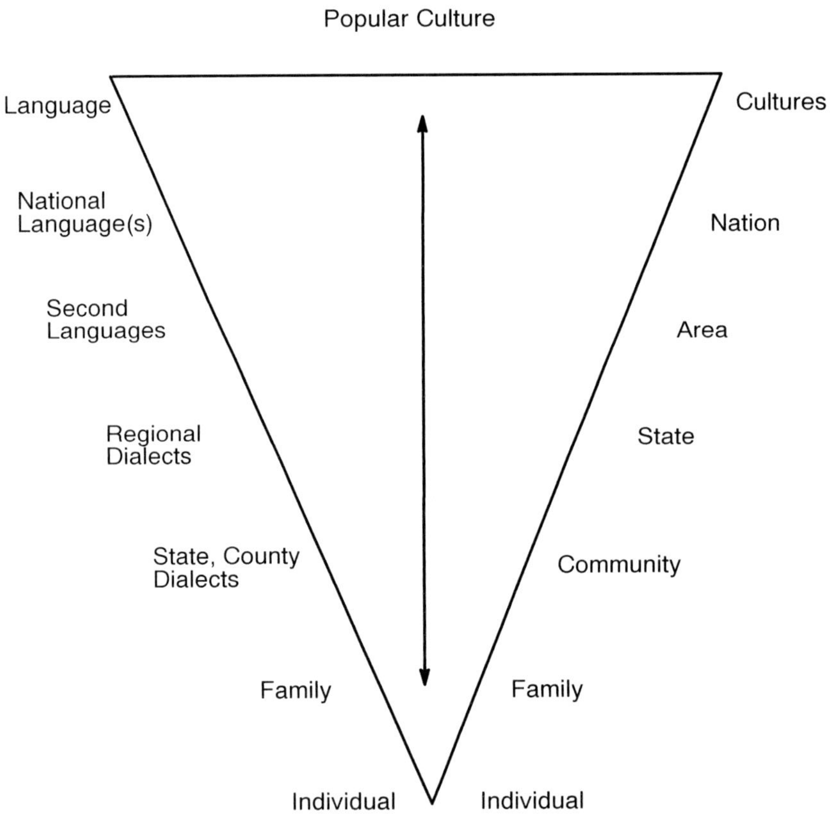

Dynamics of Origins and Developments of Culture

Failure to understand these courses, interactions, and forces of culture and cultural growth carries powerful dangers. Cultural analysts, being citizens of a democracy, often assume that citizenship and perhaps expertise in other areas automatically gives them the right and knowledge needed to become authorities in popular culture analysis, also. For example, historians, especially historians of past cultures, assume that they are already expert in popular culture and lay down the law of the dynamics of cultures of the past and the present.

But even the most perceptive scholars in such fields as history, art, literature, sociology, ethnic studies, and so on, are not automatically licensed as authorities in popular culture. The field of everyday culture, like others, such as literature, is broad and complex, and experts in some particular aspect must widen and deepen their study and understanding before they can cast more than a narrow beam of light on the whole field and before what they have to say will contain any value for the field in general. Partial or incorrect definitions and analyses are fraught with dangerous consequences because they are covered with the patina of wisdom and authority. Over 200 years ago, James Madison, known in his day as the "master builder of the Constitution," the document designed to protect the weak from the powerful, expressed his fear of the tricks of those in control: "I believe," he said, "there are more instances of the abridgment of the freedom of the people by gradual and silent encroachments of those in power than by violent and sudden usurpations" (speech in the Virginia Convention, June 16, 1788).

Perhaps even more to the point are the comments of Mike Wallace, one of the few historians who understands the dynamics of popular culture, in his book, *Mickey Mouse History and Other Essays on American Memory* (1996) when he notes that J. H. Plumb (*The Death of the Past,* 1979) said that the acquisition of the past by the ruling and possessing classes and the excluding of the masses is a widespread phenomenon through recorded time.

Wallace himself demonstrates this control by the people in power by how they presented the past, and thus the background of present-day culture in museums:

> They did so, first, by presenting particular interpretations. Of course, the museums cannot be faulted for having viewed the past selectively. There is, after all, no such thing as "the past" [or the present]. All history is a production—a deliberate

> selection, ordering, and evaluation of past events, experiences, and processes. The objection is rather that the museums incorporated selections and silences in such an order that they falsified reality and became instruments of class dominance. The museums generated conventional ways of seeing history that justified the mission of capitalists and lent a naturalism and inevitability to their authority. And, perhaps more importantly, they generated ways of not seeing. By obscuring the origins and development of capitalist society, by eradicating exploitation, racism, sexism, and class struggle from the historical record, by covering up the existence of broad-based oppositions, traditions, and popular cultures, and by rendering the majority of the population as invisible as shapers of history, the museums inhibited the capacity of visitors to imagine alternative social orders—past or future.[1]

The deliberate suppression of most of the cultural expressions of a nation is generally aided and abetted by scholars who, since they are self-styled authorities and leaders in one area of study, assume that their wisdom automatically carries over into the field of everyday culture. Because one lives in America and sings popular songs enthusiastically, sometimes one assumes that one's skill of analysis comprehends the whole culture, and guarantees a right of authority. But we all know how the archives of authority are stuffed with the bones of misassumptions and wrong analyses and conclusions. Not every scholar is a Theseus prepared to lead us through the Minotaur's maze to safety and wisdom. The floors of the maze are blocked with stacks of theories about safe routes and exits, which remind us that it is a smart person who chooses his or her guide very carefully and keeps open the sharp eye of doubt and inquiry.

Americans and America are the creators and creations of the cultures that have propelled and driven the peoples of this continent for 500 years, or more accurately for those years plus thousands preceding them. So it is not easy to answer properly the questions that have nagged the world for hundreds of years: "What is an American?" and "What is America?" Or for that matter, "What is a human being?" and "What is human society?" Answers to these questions should come from people who are more than casually interested, not from those who have arrogated an expertise to which they are not entitled.

Marshall Fishwick is a scholar and author precisely qualified to answer such questions and stimulate others with deep knowledge and keen insight. Through the years he has analyzed these questions closely and clearly. He began as a poet, thus able to see beneath the surface of reality and understand the spirit of life. He supplemented his Yale degree in American studies by continuously feeding his curiosity through reading, listening to, and participating in all aspects of literature, popular culture, and life.

Throughout his forty-year career, Marshall has been interested in the dynamics of popular culture, not only in the moving parts but, more comprehensively, in the organism as a whole. He has always specialized in generalities, ranging from the particulars to the whole. This is contrary to the general academic activity which, unable to read the whole, has concentrated on the particular, filling a warehouse of scholarship with well-polished parts but no schemata for assembly.

But Marshall has always wanted to communicate with his fellow citizens as well as fellow scholars. For years now he has been an occasional contributor to newspapers and magazines and the electronic media. In this volume, he wants to talk to people on all levels of the communication scale, scholars, students, and other citizens, that is, to Everywoman, Everyman, and, at least through them, Everychild.

Marshall's interests and sweep are comprehensive and deep, beginning with the first communication in popular culture when one Homo sapiens said to another for the first time, "Hey, have you heard . . . ?" down to the latest communication from any of the various media that swirl around us all the time. In other words, Marshall, as an American and as a citizen of the world, wants to talk to all Americans and citizens of the world.

To do so, one must know whereof one speaks. Marshall does. He has been there, done that. And he has analyzed what he did and why he did it. Now that produces a useful script for understanding culture, a readable and clear map useful not only in helping one get to a destination but also in explaining the phenomena encountered along the way and the meaning of a destination once one has gotten there.

Marshall is just as much at home in cavespace as in cyberspace. From the starting point to our present-day site is a great distance in time and cultural sophistication. Marshall has examined the roadways from then to now and the watering holes at which for one reason or another we have paused. He has looked down every avenue, up every height, heeded every expression, and probed every question he has encountered on his trip.

We need look no further for an enlightened and sparkling guide to help us along the way.

Ray Browne
Secretary-Treasurer, Popular Culture Association/
American Culture Association, and Chair,
Popular Culture Department, Bowling Green University

Preface

Marshall Fishwick is one of the leading scholars on earth when it comes to the study of popular culture—indeed, with Ray Browne, Russel Nye, and Tom Towers, Marshall Fishwick founded the Popular Culture Association in 1968 as a breakaway group that defied the so-called "wisdom" of the academy by declaring that popular culture studies must become part of the humanities.

At least in private, Americans will admit that they have learned as much about life from popular culture as they have from their families, teachers, religious leaders, and peers. Popular music not only surrounds us in our homes, but also pursues us to work in our automobiles; after work, it soothes and prods us as we shop. Television shapes our understanding of the political and emotional universes; of course, while we watch the news, we absorb countless commercial messages. With the proliferation of video stores, we have increased access to images of men and women from an enormous pool of Hollywood productions—going back to the turn of the century and progressing forward to the latest versions of *Die Hard* and *Lethal Weapon*. At civic clubs and churches, we chat about the most recent John Grisham novel or share our fascination with Susan Howatch fictions such as *Glittering Images* and *Glamorous Powers*—the latter are especially popular in Episcopalia, a place Fishwick visits each Sunday morning.

Back in the late 1960s, Marshall Fishwick and his fellow pioneers encountered resistance to their work. But over time, the popular culture studies movement has become a veritable stampede. Check the advertisements for nonfiction books in *The New York Times Book Review* and you will discover that every university press—indeed, every commercial press—is now zealous to attract readers with insights into popular culture. We can trace this revolution back to Marshall Fishwick and his colleagues.

Unfortunately, it is often the case that scholars study what is popular by applying arcane methods. During the 1970s, the theory of semiotics took hold in the profession. Later, Freudian (perhaps better labeled "pseudo-Freudian") models were applied to prove the irrationality of a "mass" public. Today, gender, class, and "gaze" theories seem to be popular. Marshall Fishwick's work takes giant strides over such methodological tanglefoot.

Indeed, *Popular Culture: Cavespace to Cyberspace*—the book in hand—ponders the origins, history, and impact of popular culture from the earliest days of human history. Refreshingly, it does so in ordinary language. As a senior scholar with over twenty books to his credit, Fishwick is secure enough to speak in a way that his fellow citizens can understand—plainly and effectively—about the impact and meaning of the signals we encounter each day as we move through a complex information environment.

One of the aspects of popular culture studies—as opposed to the study of what is called "pop culture"—is a love for democracy. Both Marshall Fishwick and Ray Browne committed themselves to the ideals of democracy in their youth by serving in the military during World War II. Just prior to the D-Day invasion (brought back to the world so vividly in 1998 by Stephen Spielberg's *Saving Private Ryan*) Lieutenant Marshall Fishwick, serving aboard a seagoing tug, towed a portion of Churchill's famous prefabricated concrete "Mulberry Harbor" (an artificial harbor constructed for Normandy beach) across the Atlantic Ocean from Rhode Island to Southhampton. Later in the war, Sergeant Ray Browne reported for duty in Stuttgart, Germany. Both of these pioneers love America and honor ordinary people and their tastes. The work of the Popular Culture Association, the American Culture Association, and their journals reflect this faith in the democracy. Without such *Verständnis* (or sympathy), popular culture studies become a cynical unmasking of the manipulation of masses by elites—the kind of attitude found in Barry Levinson's film *Wag the Dog* (1997). (Contrast Levinson's cynicism with the dignity of ordinary people in Stephen Spielberg's World War II blockbuster.) Marshall Fishwick is no castigator of popular tastes; instead he explains how the popular mind seeks out heroes, myths, images of the good life.

Yet Fishwick remembers how difficult it was to conduct such studies in the early days: "Graduate work wasn't designed to erase the boundaries—but to bolster the elitist walls. The higher the holier—the closer you get to your PhD, the deeper the seeds of elitism are planted. One could go and enjoy a 'Hollywood flick,' or enjoy an outing with 'the great American people,' but must retreat to the stacks and seminars as quickly as possible." During a Fulbright fellowship to Borneo, Fishwick was confronted by questions concerning the deep (and popular) roots of culture—questions that led him to reconsider the elitism of his Ivy League education. (Margaret Mead had been inspired to ask similar questions during her visits to Borneo in the 1930s.) As he says with the excitement of a convert to holistic studies of culture, "I suffered from culture shock in a distant wilderness. But in a flash I came to realize that America, too, is a wilderness shared."

Popular Culture: Cavespace to Cyberspace is the distillation of Fishwick's findings. It is written in the style of his newspaper and magazine articles over the past thirty years—in a language that is evocative rather than academic, full of metaphors that reflect a sprightly imagination. This book will help readers to understand at least eleven major challenges in our contemporary cultural and political life:

1. How do we become attracted to issues, leaders, and causes? What is the power of presidential symbolism in our national life? How do media images function to assist us (or obstruct us) in our roles as citizens?
2. Why do certain symbols emerge to represent our deepest feelings as a people?
3. Time is so important in our culture; how did the obsession evolve?
4. How do buildings express our values? What are some key examples of this physical objectification from the Greek, Roman, medieval, and modern eras?
5. Are our modern myths to be found in the works of Walt Disney and other Hollywood visionaries?
6. How do our war memorials and our memorials to presidents express the American spirit?
7. Elvis is a lasting image of popular culture. How do we better understand our fellow citizens and ourselves by exploring the

meaning of his image? Why do so many thoughtful people attempt to deny the obvious popularity of "The King"?

8. Anthropologists have studied the relationship of dance to culture for decades. What about America's dances over the years?
9. And what of the Internet? How do the lessons of popular culture in the past prepare us to cope with the challenges of a virtual future?
10. Given the outward thrust of American music, films, clothes, and fads, what is the future of world cultures? Should we fear the Americanization of the globe? Can East meet West in a way that will meld the best of both cultural traditions?
11. What is the future of faith? Have secular rituals—such as the Superbowl—replaced traditional symbols in our national life, leaving religion by the sideline?

Our author promises that there is hope for us, despite these challenges and questions. At the end of history, he exhorts us: "Thinkers of the world unite—you have nothing to lose but your silicon chips." We need to understand our world and transform it into a better place: "We are the system—we accept and abide by its dictates. If there is an enemy, it's us. We repeat and endorse platitudes and truisms. If and when we really want to, we can change them." Once we appreciate our involvement and need for popular culture, we can define a better future: "We must find both mission and method by which to reach a new consensus and understanding . . . construct a new and better world. We did it before. We can do it again."

As the new millennium approaches, Marshall Fishwick's panorama of the popular will give us perspective, understanding, and energy. Hopefully, we will emulate this author's enthusiasm for culture (both popular and elite)—and for life.

Peter C. Rollins, PhD
Regents Professor of English and American/Film Studies
Oklahoma State University

What Is Our Goal?

This book is meant as a textbook and trade book, not a scholarly monograph. What I seek to "trade" are observations and ideas coming from a life of teaching and writing about popular culture. The target is the general reader rather than the specialist. There are a few reference notes, but I have usually preferred endnotes, suggesting what sources I have found most helpful, and which can lead the interested reader into deeper waters. An extensive list of books for further reading can be found at the end.

I want to show how popular culture has dealt with the perennially great questions: good versus evil, truth versus deception, life versus death, reality versus illusion. Whirl has always been king, and the people have always had to contend with baffling change. They have faced wars and rumors of wars; floods and famines; darkness and despair. Yet they have not only survived—they have moved forward to new and greater epochs.

All sorts of questions have arisen, and some remain. I shall attempt to raise those questions anew, and to persuade readers to ask questions for themselves. In this sense, our goal was best stated by Socrates: the unexamined life is not worth living.

* * *

PART I: FORETHOUGHTS

Back to the Ice Age

On September 19, 1991, two German mountain climbers, Erika and Helmut Simon, made a remarkable discovery in the lofty Dolomites between Italy and Austria. In a deep gulch on the edge of the Simulaun Glacier, which has partly melted in recent years, they came upon a completely intact mummy, still partly covered by ice. On occasion, soldiers who died in fierce fighting in these mountains during World War I have been uncovered, along with their bombs, shells, hats, and shoes. Glaciers are retreating because of hotter temperatures. The hikers assumed this must be yet another wartime victim, who had perished and frozen, only to be discovered in warmer weather much later.

Both the Italian carabinieri and the Austrian polizei were called to the scene. No one was sure just where the border between the two countries ran. The Italians thought the mummy had come from the Austrian side, so the body was taken down Austria's Oetztal Valley to Innsbruck.

Scientists at Innsbruck University examined and analyzed the remains, and made an astonishing report. They were 5,300 years old. The hikers had discovered the best-preserved body of an ancient man ever uncovered. He had died over 3,000 years before the birth of Christ, when the Egyptian pharaohs were building their pyramids.

Believed to be a hunter about forty-five years old, he carried a spear and a bronze ax, and wore boots with dried grass inside for warmth, fur leggings, bearskin booties, goatskin undergarments, and a grass-woven sleeveless cloak, open down the front. They had the whole man, dressed and armed in the Ice Age. Scientists believed he fell asleep in the bitter cold and froze to death. Perhaps the fact that he had been in total darkness over the millennia explained the remarkable preservation. The organic components of his clothing and equipment have never before been found in such an excellent state.

The news electrified the world of archaeology. Now Italy wanted the treasure, claiming it was found on Italian territory. A joint commission decided the body had indeed been discovered a few yards inside the Italian border. Judging from his equipment and the food from his last meal, the man had come from the south. He belonged to Italy—and to a new mythology.

How had this primitive warrior managed to climb to the frozen heights of the peaks? What was he doing there? What was he hunting? He was quickly linked with a local myth of popular culture and the mountaineers who fought for freedom and traditional Roman Catholicism against Napoleon and Bavaria, inspired by the Holy Heart of Jesus—the traditional ethnic symbol of their devotion.

The Iceman needed a nickname for newspaper headlines, tourists, and promoters, and got one: Oetzi (or Ötzi in German) and so he has been ever since. Oetzi the Iceman! I could hardly believe my good fortune. Here was a largely intact Bronze Age man being brought in a single leap into the late twentieth century—from cavespace to cyberspace—soon to become a tourist attraction and popular hero. The book I was already writing, connecting the prehistoric and the postmodern, suddenly had a symbolic home in the Dolomites—a local habitation and a name.

Reluctantly the Austrians agreed to the return. Threats from Austrian nationalists to intervene prompted Austrian police to escort Oetzi through the Brenner Pass to the border, where Italian police took over. Both Italian and South Tyrolian flags flew on the return route. The corpse would find a home as the centerpiece of the splendid new South Tyrol Museum of Archaeology in Bolzano, which rejoiced at their new arrival. "We have a candid snapshot of daily life more than 5,000 years ago," said Lorenzo Dal Ri, museum director. "We have a unique treasure."

The Iceman's ax shows that more than 5,000 years ago metalworking had already reached an impressive level of development. Looking at the Iceman, we focus on the everyday material culture of his time and the level of technological development.

The quiet town of Bolzano, known for its fine buildings, a highly rated restaurant (The Lauren, named for the mythical King of the Dolomites), and peasants who still wear their blue aprons while working in the fields, was about to be thrust into the Cyberworld of Celebrity. A new state-of-the-art museum was built to house Oetzi, with a specially designed refrigeration chamber to preserve him for generations to come. Once the museum opened, 10,000 people visited Oetzi in a few days. Exposed like the Holy Shroud of Turin, he was in a bulletproof casket with special inside lighting. Viewers looked into the dark room through a small window, to pass by, observe, and marvel. The ancient hunter who died in the ice at the top of the mountain had entered a whole new life. He had become a new pop icon, and a world-class tourist attraction, like King Tut.

In an age of consumer capitalism, entrepreneurs rushed forward. Local businesses, cafes, shops, and hotels prepared for the new day. "I didn't know what to do," said Elmar Streitberger, owner of a family-run cafe a few steps from the archaeology museum. "After all, Oetzi is a dead man."

But tourists are alive and hungry. Inventive Elmar came up with a novel design—Oetzi's head and arms made out of nuts and chocolate, with a yellow candy hat on top. He had to work fifteen or sixteen hours a day to prepare for the tourists, but he did it. Other locals rallied too. Soon Bolzano pilgrims could buy Oetzi pizza, Oetzi ice cream, and a variety of souvenirs. Outside the local coffee shop a poster advertised a drink called "Der Mann im Eis" (The Iceman).

An ice cream parlor mounted a colorful sign showing two white mountain peaks against a blue background, with a tiny man (presumably Oetzi) climbing one of the peaks. Inside one could purchase Oetzi ice cream—vanilla with caramel and pine nuts. Down the street a bakery has created an Oetzi original—a milk chocolate figure lying on a bed of pure white chocolate. There is also Oetzi wine, with a watercolor of the mummy on the label, and a series of colorful postcards.

Would the real Oetzi disappear in a sea of similar contrivances? The head of the museum, Dal Ri, fought back. "We've hired a lawyer to study whether we can copyright Oetzi," he told the press, "but I fear we've waited too long, because it's been seven years since the name was coined." The case is pending.

Meanwhile, Oetzi himself has undergone an incredible transformation—from the timeless freedom of the white glaciers to a glass cage, and to the prism of the cyberworld. Now he is a world attraction, a kind of immortal image of death. He seems like a creature coming from a faraway world, a UFO from outer space. Has all our new technology and dependence on virtual reality transformed the old hunter into a UFO Man?

Oetzi becomes a kind of heroic metaphor for this book—the ancient Iceman who has suddenly become a postmodern pop icon. Can and will he help us understand the role popular culture has played in the intervening 5,300 years? A daunting task. Let us leave Oetzi in his unique resting place and begin our long journey.

Definitions

This is a book about popular—that is, people's—culture. "Popular" is derived from the Latin *populus,* meaning people. As long as there have been people, they have had their stories, their struggles, their cultures. But what people? Where? How can we tell their story going back to countless centuries when history was blind?

Of the people living before recorded history, we know almost nothing; and of those who lived in the historic past we know very little. Histories of civilizations tend to concentrate on political and military leaders, and thinkers. Who speaks for workers, women, slaves, criminals, outcasts, and peasants? For the inarticulate, the ignored, the oppressed? Great revolutions thrust them forward. We examine "the people" in slave revolts and the American, French, Russian, and Chinese revolutions, and acknowledge the enormous power of Marxism with its stress on the proletariat. But hasn't it collapsed as our century ends? Can popular culture become unpopular?

Who can say how many definitions of popular culture exist, and what they have in common? We know there are references to the rabble, masses, hoi polloi dating back to Plato's time; but when were "popular" and "culture" linked? The Greeks had no word for "culture," but the Romans spoke of vox populi—"the voice of the people." The first documented modern reference for "culture" dates from 1516 and the first for *popular* culture (more specifically, *Kultur des Volkes*) by Germany's Johann Gottfried von Herder in 1781. He is a major figure in our story, and deserves more credit than has been given him.

Born in East Prussia in 1744, educated in Königsberg, Herder went to Riga in 1764, and from there made a long sea voyage that changed his life. He discovered a vibrant folk culture to which he devoted the rest of his life. On this voyage he saw himself as having left the safe shore of elite culture behind, and as having come to understand the meaning of feeling *(Gefühl)*. He saw poetry as the mode of expressing this feeling, arising from the natural environment and the people's inheritance, and the poet as the creator of the nation.

He found the greatest and purest poetry in the "uncivilized periods"—the times of the Old Testament, the Eddas, Homer, and Ossian. A main creator of the *Sturm und Drang* movement, he believed that nature and history obey a uniform system of laws. He was an incredibly prolific author, as thirty-three volumes of Herder's work were published between 1805 and 1820.

In a series of letters and articles he raised crucial questions: How can we understand people who neither read nor wrote, and left no records? Were elite informants able to "get into" peasants' minds, relate to their feelings? Might they have been misled or deceived—as in the famous twentieth-century case of anthropologist Margaret Mead in her highly touted book, *Coming of Age in Samoa* (1928)?

If we have a single text for an important topic—such as seventeenth-century Russian tales and folk songs, collected by two British visitors (Richard James and Samuel Collins)—can we assume they are "authentic"? What about popular events known only through "the authorities" who suppressed them? We only learn about heretics and witches, for example, through their prosecutors. People burned at the stake or hanged do not get to give their rebuttals.

Can we really understand active performance (such as festivals, dramas, or sermons) if we have only words, not action, to guide us? Who chooses those words?

Popular culture is both an elusive and an unpredictable quarry. No one knows when an unexpected event will capture the world's attention: when the assassin's bullet will strike or when the ship will sink, causing an outburst of popular interest.

In March 1993 the outburst centered on a seventy-eight-acre compound outside Waco, Texas, occupied by the Branch Davidian cult. FBI agents staged a surprise attack in an attempt to arrest cult leader David Koresh on weapons charges. The cult members—prepared and perhaps forewarned—killed four of the federal agents and wounded another sixteen. Koresh refused to surrender, vowing to hold out until God told him what to do. He waited too long.

Eventually the nervous government agents crashed through his stronghold with tanks, set fire to the building, and waited for all inside to emerge. Many did not emerge—they burned to death.

Charges flew back and forth. Shortly before that, New York's World Trade Center had been rocked by a devastating explosion, killing six people and injuring hundreds. Worse was yet to come. In 1995 the federal building in Oklahoma City was leveled by a huge explosion that killed 169 people and injured hundreds more. Was terrorism a new form of popular revolution?

It was a global question. Islamic fundamentalists in Egypt and Algeria went on rampages. No one could quell the violence between Catholics and Protestants in Ireland. Yugoslavia fell apart while a helpless Europe watched. The Orthodox Serbs, Islamic Bosnians, and Roman Catholic Croats all waged a new crusading warfare, even genocide. Respect for religion in areas of popular culture such as music, books, and the Internet reached a new low.

The death of Britain's Princess Diana in an automobile speeding through a Paris traffic tunnel turned out to be, in some quarters, the event of the year in 1997. The whole world went into shock and mourning, and many thought the British monarchy might collapse. It didn't.

How much was happening? How much was hype? Does our media allow us to think that what we see on the tube and the screen is "reality"?

Teach us not to live by "the polls," conducted by who can say, answered by who knows? The phrase "tabloid journalism" has lost its meaning—isn't much of it tabloid? Will disinformation push information out of our lives? Will we let the motto of our local TV news station—"If it bleeds, it leads"—become our new national motto?

Good-bye to any simplistic definition of popular culture. It is best thought of metaphorically—a wandering minstrel, a thing of shreds and patches, for example. Or a kaleidoscope, which changes constantly as we twist it in our hands. Our new electronic Proteus, taking on forms and shapes of all things on land and sea.

"Different strokes for different folks." In the Old World, thoughts turn to the archaic and traditional; in the New World, to the experimental and electronic. The French honor the rural tradition of their *paysans*, the Marxists, the shattered dream of rule by the proletariat. We have a different outlook in the United States. We are go-for-broke, get-it-all-now consumers.

There is still a problem. We must have the car of tomorrow before we've paid for the car of today. Who would risk the slings and arrows of outrageous fortune by being behind the times? Thank God for credit cards.

Marshall McLuhan was right: if it works, it's obsolete.

Over a hundred years ago, Walt Whitman noted that "culture" is the word of the modern. Though few could agree on just what the word meant, everyone thought it basic and essential to our "modern way of life." Whitman never used the term "popular culture" even though he came to personify it.

At that time, the literati still followed the banner of Matthew Arnold, who believed in the pursuit of perfection—a "high" calling, involving beauty and intelligence or sweetness and light. Newly empowered social scientists wanted culture to reflect the multiplicity of human behavior in different places and times. The aging robber baron J. P. Morgan had a simpler definition: Culture is something that comes when people have more than two servants. More recently, Herbert Read, the British art critic, gave up on the squabble, titling his new book, *To Hell with Culture*.

Whatever the definition, the cultural ideal really meant the imperializing domination of some national culture, such as the American way of life, British hegemony, *Deutsche Kultur*, or the Communist International. None of these patriotic definitions describe the "culture" of the Global Village.

We need a new way to present and understand material that is being labeled "postmodern"; a larger framework or stage, on which to depict twentieth-century culture moving into the twenty-first century. New wine in old bottles? Or might the bottles themselves need to be remade and relabeled?

Recent books search for new formats. New journalism purports to be more subjective, personal, combative. New historians stress thematic rather than chronological lines. Others recast the whole past in light of a new ideology, such as feminism, multiculturalism, sexual liberation, or deconstruction. Two examples are Hugh Thomas's *World History: The Story of Mankind from Prehistory to the Present* (1996) and Theodore Zeldin's *An Intimate History of Humanity* (1995). Zeldin tells us just what he hopes to do—to show how, today, it is possible for individuals to "form a fresh view of their own." Various fresh views flood the market.

Hugh Thomas promises a personal view, and wants to encompass humanity's "whole record." Both include healthy doses of popular culture and personal speculation. No more strict "scholarly objectivity" for them. The era of DWEMs—Dead White European Males—has ended.

Or has it? Certainly neither Thomas nor Zeldin seem dead or even burned out. Once again some of the best art and literature of our time comes from live Europeans—some of them even white males.

In popular culture we hesitate to pronounce premature deaths. Dead men (such as Elvis) do tell tales. Nostalgia and revivalism are powerful forces. Discarded TV series (such as *The Brady Bunch* and *I Love Lucy*) become more popular than ever. As I write this, a whole series of Tarzan movies captivates a new generation.

"Radical" and "exclusive" groups mellow and modify. The new black studies and women's studies scholars have new agendas. Sometimes strongly ideological groups end up sounding just like the things they condemn. The John Birchers, for example, became as close-minded and dogmatic as the Communists. "Swing votes" swing wildly in courts and legislatures, and politicians vote for things they promised to oppose. A leader is on the wrong side of a "burning issue." "The crowd has turned around," he laments. "I must rush and get in front so I can lead them."

Where will "the crowd" go next? Who can say?

ENDNOTE

The much-heralded worldwide "popular culture explosion" has produced thousands of books, films, TV programs, and monographs but no widely accepted definition. By best estimates over a million students now study popular culture in America. Who can say how many paths they pursue, how many topics they cover? And how does the subject fare abroad?

If there is a central clearinghouse, it is the Popular Culture Association and Popular Press, both located at Bowling Green State University, Bowling Green, Ohio 43403. Their journals and press continue to produce new material, book reviews, and conference notes. Catalogs and journals are available on request.

The Arno Press, a New York Times company, produced an ambitious twenty-six-volume series, *Popular Culture in America,* during the 1970s. My own best efforts to define this elusive and pervasive field are *Parameters of Popular Culture* (Bowling Green, OH, Popular Press, 1975) and *An American Mosaic: Refiguring American Culture History* (New York, American Heritage, 1997).

Popular-Common Culture

To cope with our large undertaking—to make an aerial mosaic of "the people" that stretches from prehistoric to postmodern times—we shall use two terms interchangeably—"popular" and "common." They are more inclusive than the traditional academic terms, and will give us a broad foundation on which we can build a structure with many rooms. These are not easy concepts to pin down.

The origin of the word "common" is uncertain. There are countless references and many meanings. "Common" things are bound together, shared. They make our lives significant and symbolic; they make us human.

The great Greek philosophers, such as Thales, Socrates, Plato, and Aristotle, developed this idea. So did the pivotal early thinkers of India, China, and Egypt. The Christians seek common bonds and meaning in both the Old and New Testaments. The Roman Empire centered around a common language, common law, and common purpose. The true secret of Rome's greatness and longevity is that it sustained variety but insisted on cohesion. They utilized the media (plural of the Latin word *medium*—intermediate or between) as well as their legions. For centuries Latin was the predominant medium in the West, as English has become today. Media makes information, cooperation, and popular culture possible.

No one comprehended this more clearly than Marcus Tullius Cicero (106-43 B.C.). If popular-common culture has a godfather, it is this remarkable man. He spent years studying, selecting, summing up, so that he might transmit great works from other languages (especially Greek) into easily available Latin. He knew that being simple is never simple, and that being widely understood is the highest possible achievement. He helped make Latin into a language of believable ideas and wide import. He brought over from Greece a way of thinking—the dialectical—and made possible what we now call the academic way of life.

One idea Cicero absorbed from the Greeks was that of a sound mind in a sound body. All parts must function in harmony with the others, and the whole must reflect and stay in phase with nature. Exercise the muscles to keep the mind alert, and the body in harmony with the mind. Rejoice in sport, play, and merriment. The Greeks and Romans did this—and so do we. Are not the Super Bowl, the World Cup, CNN, and the World Series worldwide symbols that tend to link all of us together in a single chain?

Such universal ideals lived on into the Middle Ages; Aristotle had great influence on the church fathers, such as Saint Augustine and Thomas Aquinas. Aquinas, in particular, understood and endorsed what we are calling common or popular culture. "All who are included in a community," he wrote in his *Summa Theologiae*, "stand in relation to that community as part to the whole." All nations, regions, and institutions come to know that only by creating community can they survive. True community respects the individual, whatever that person's rank or station; it is immediate and real. False community centers on groups or ideologies; it tends to be abstract, remote, and artificial. The end product of true community is concern; that of false community is exploitation.

The community shares artifacts, mentifacts, icons, ideas, and yearnings. Language, codes, gestures, rituals, and symbols are means of expression, but the essence of community—and hence of popular culture—is intangible. It goes under such names as the Holy Ghost, the Great Spirit, *Volksgeist*, the general will. "Behold, I give you a mystery."

It is difficult for us to grasp what common culture was like before the Industrial Revolution. What did the common people do? We must think away television, radio, cinema, computers; universal education, universal literacy, class consciousness, nationalism. We must put aside our confidence in science, technology, and progress. To the extent that we can reenter this now-distant world, we will understand that it had two separate tracks: the great tradition and the little tradition. The great tradition was for the few who attended schools and universities; the little tradition included all the rest—the nonlearned, the unlettered, the nonelite.

Hence the great tradition was closed, in that people not privileged to advanced education couldn't share it. Those outside quite literally "didn't speak the language." The little tradition, transferred informally by word of mouth, was open to all. It was, in the basic meaning of the term, popular as well as common. The educated elite, having access to both traditions, were bilingual and bicultural. The great tradition was serious. The little tradition, at home in tavern, fair, and marketplace, was often playful.

The great tradition included what we now call the classical tradition, with its philosophy and theology, as well as later movements—the Renaissance, the Scientific Revolution, and the Enlightenment, for example. The little tradition clung to the things it knew best—folktales, folk songs, mystery plays, fairs, and festivals. For centuries their great days were Christmas, New Year, Carnival, May Day, and Midsummer. Some of that greatness still remains.

There was never one unified great or little tradition. They always differed according to age, wealth, religion, and gender. The great tradition provided the rulers, scholars, writers. The little tradition provided craftsmen and peasants; shepherds, sailors, thieves, and beggars. Our best accounts of all this are Peter Burke's *Popular Culture in Early Modern Europe* (1978), and Robert Redfield's *Peasant Society and Culture* (1967).

The daily life and culture of the prehistoric and preindustrial world is elusive. The literate modern historian cannot hope to understand people unlike himself or herself; their attitudes and values; their way of thinking and acting. All we can do is gather as much evidence as we can and try to recreate aspects of this vanished world.

What we can document and study in detail is the splintering of common (or popular) culture. A whole series of revolutions (political, religious, social, economic) combined to give us what we call the modern world. Power was consolidated. The peasant was pulled into the cities, and became the factory hand, the proletariat. There were changes not only in clothing and currency, but in sentiment and mind-set. New nations, creeds, and ideologies sprang into existence.

Starting with James Macpherson's *Fragments of Ancient Poetry* (1760) and *Fingal* (1762), writers and historians tried to document the fast-disappearing prerevolutionary world—Thomas Percy, Samuel Johnson, V. F. Trutovsky, J. G. von Herder, Sir Walter Scott, and the

Grimm brothers, for example. As the gap between the two cultures widened, some educated people began to collect and study popular songs, beliefs, and festivals. There was a remarkable change of attitude among the educated. In 1500, they looked down on the common people, but shared their culture. By 1800 their descendants no longer participated in what had become popular culture, but they strove to rediscover it. They would even come, in due time, to admire "the people," and to pursue their fast-disappearing legacy. We are following in their footsteps.

* * *

On several occasions "Everyman," used in the generic sense, pops up in our story. Please don't think of this term as merely male. It represents everyone—male, female, old, young, white, non-white—a stick figure standing for "all of us." As a species we are all one.

Most people meet Everyman in John Bunyan's *Pilgrim's Progress*. What a wonderful title! We are all pilgrims, working our way through space and time—going somewhere, but who can say where? All ending up in the same place. "The paths of glory lead but to the grave."

Bunyan (1628-1688) was "one of the people," lived a hard life, and is a good patron for a book on popular culture. We can profit from reading Bunyan, but don't want to end up as he did—troubled by a maniacal impulse that prompted him to pray to the trees, to a broomstick, even to the parish bull. His great book, published in 1678, went through ten editions by 1685. How few of us can hope for such a readership today! Still, as pilgrims, we plod forward.

A Movable Feast

To say something is or is not popular raises several questions: when, where, for how long? Popularity does not offer a permanent banquet, with assured seats for its favorites. Instead, it is a movable feast.

One example will make the point: When Will Shakespeare's plays were performed in the Globe Theatre, they were "popular," enjoyed by ordinary people who laughed, grunted, and snickered at them. They must have had many words for the males who played female parts, and for the "mechanics" who mirrored working-class people of that time.

Over the years Will became William, and his plays became so "highbrow" that only the privileged and educated could, we were told, "appreciate" them. Both undergraduate and graduate Shakespeare courses cropped up in universities. Students got PhD degrees for dealing with small segments of a single play, or for finding interior meanings that might have startled Shakespeare.

With the advent of film production of Shakespeare all this changed. Producers found out that the public—including those without PhDs—loved Shakespeare and flocked to see his work on stage and screen. Shakespeare festivals flourished, and actors such as Laurence Olivier and Mel Gibson, directors such as Franco Zeffirelli and Baz Luhrman showed just how popular Shakespeare had become—four centuries after many of the plays were first produced. At one time in the mid-1990s five Shakespeare plays were either being planned or in production in Hollywood. What living playwright could match that?

Who can deal with the new world in constant crisis, both imploding and exploding; babies gurgling and AIDS victims dying; new crises and stalemates in Washington; the yells of tens of thousands at the Super Bowl, and the chirp of a single cricket in an unending Kansas wheat field. All these are part of postmodern America.

We live in a place and time that thrives on opposites. It soothes and irritates, understates and overstates, inflates and deflates. The stock market goes up and down like a roller coaster. Long on compression, we seem short on compassion. That compression, epitomized today in the merger, downsizing, cartoon, and sound bite, has had many names over the ages: the quip, retort, one-liner, epigram, bon mot. The list of the historic master compressors is impressive: Buddha, Aesop, Jesus, Paul, Petrarch, Pascal, Lincoln, Will Rogers, to name a few. And don't forget Winston Churchill, John Kennedy, Margaret Thatcher, Johnny Carson, and Roseanne. They all knew how to reach the hearts and minds of many.

Popular culture is a mirror held up to life. Sometimes that reflection seems full of motion and madness. Repeating old meanings, it confronts new factors—physical and social mobility, global revolution, the collapse of old ideologies, and the meteoric rise of new technologies, to name a few. In trying to define it we get no fixed boundaries but changing spaces—Einstein's answer to Descartes. With popular culture, relativity is all. No wonder it has trouble becoming a "discipline" in an unpredictable and undisciplined world.

The great mystery is the audience. What happens when an actor or musician performs? Why do some succeed, others fail? Examining scripts, songs, views gives few clues. The crucial element is the audience—the great unknown. The chemistry involved is complex and unpredictable.

Why are certain audiences especially warm and appreciative? Can there be great performances without great audiences? What has mass media (in which audiences are completely isolated and passive) done to the whole concept of audience?

What about the Common Man—alias the Slob? Is he no better than Caliban, caught in the ooze at the bottom of the mass mind?

There have always been those (José Ortega y Gasset, Dwight Macdonald, Ezra Pound, R. F. Blackmuir, for example) who insist that this is so. "If the 243 Americans who ever heard of civilization would stop crabbing each other and organize," Pound said, "that would be a start."

The 243 have not prevailed. Instead, writers and scholars, working in different fields and nations, have turned the tables on the snobs and elitists. No one ideology or strategy in writing history is "nobler" or "better" than another. All cultural data have the same value as evidence of a culture's fundamental form. The people we have called "common" or "primitive" are complex, sensitive, and adaptable like any other group.

I speak of structuralists such as Claude Lévi-Strauss and Jean Piaget; theorists including Noam Chomsky, Northrop Frye, and Marshall McLuhan; critics such as Paul de Man and Geoffrey Hartman; social scientists including Hayden White and Dell Humes. Metahistory has been successful in stressing and documenting the potential equality of all humankind.

Andrew Levison's book *The Working Class Majority* (1974) suggests that many widely held stereotypes are not true. Careful studies indicate, for example, that the working class is not more conservative and racist than other classes.

Perhaps Thomas Jefferson was right when (responding to the French aristocrat Du Pont) he wrote that if you give the people light they will find their way. Horatio Greenough echoed the same thought, writing in the *Democratic Review* (1843): "It is the great multitude for whom all really great things are done, said, suffered. The multitude desires the best of everything; and in the long run is the best judge of it."

Greenough was using the media—in this case, the printing press—to make critical comments. Later, electronic media provided voices for other critics of all persuasions. With the invention of xerography, everyone became a publisher. Photocopying not only duplicates print, but changes the way we deal with it. To categorize, then duplicate, has become part of our expectations.

If the media criticizes the culture, who criticizes the media? This has become a crucial question along with the continuing demand for more openness. Poking fun at pompous media lords goes back to *Puck* and *Judge* in the 1890s, and continues with *The New Yorker* and *National Lampoon* in our day. But who dares tackle the Associated Press, that Max Eastman called "the worst monopoly"? And who serves the function which Sweden recognized in 1969 when they appointed an official ombudsman, "to serve as a commissioner of grievances and to prosecute violations of press ethics"?

Will the people themselves perform this task? There are signs of "demassification" in the rise of ethnic consciousness, secessionist movements, and regional and local publications. Perhaps a revolutionary new information system is being born.

Ecology has become a mass movement. First coined by Ernst Haeckel in 1869 (from the Greek *oikos*, household), the term "ecology" was long used mainly by scientists. But the idea of preserving the balance of nature has seized the public imagination; the human-media relationship is changing accordingly.

As we enter a postmodern world—with a postindustrial society—we will have to ask questions that we once considered answered:

1. What do we mean now by "cultural innovation"? How does a term that was mechanistic apply to an electronic age?
2. How have electronic mass media changed the arts? Morality? Education?
3. What happened to the old "publics," and who are the new ones? Who are our new culture consumers?
4. Is there such a thing as postmodernism? What's behind the label?

These are earthbound questions; we have moved into space. When the astronauts transmitted back to us pictures of the earth as a small orb floating in a sea of space, everything altered.

En route to the moon, we perceived for the first time what it meant to be an earthling. What new perceptions will we have before the twentieth century ends? What will our popular culture be like in 2020 A.D.?

Too many factors and actions refute the charge that there is no purpose, elasticity, or individuality left in modern society. True, our world supports many monopolies and conformists, but often their stripes are only skin deep. The problem is to see *underneath* the surface—of our society, our technology, our kitsch—and identify new wellsprings of energy, technique, and faith. This will mean clearing not only the junkyards from our highways, but the bric-a-brac from our intellectual attics. We must civilize technopolis, where barbarism and anonymity are well entrenched. To do this, we must comprehend our environment in new ways. Weeding and watering the landscape of the mind, we may find an America that looks ever less like a Currier and Ives landscape and ever more like a

Jackson Pollock abstraction. Other epochs and leaders await their turn. The question raised by William Butler Yeats in "The Second Coming" remains unanswered:

> And what rough beast, its hour come round at last,
> Slouches towards Bethlehem to be born?

Perspectives

Pooh Bah, Lord High Everything in *The Mikado*, was born sneering. He could trace his ancestry back to a protoplasmal primordial globule. Not me. I was told never to ask questions about my grandfather, and never even to mention my Uncle Willie. Nevertheless, I must confess: I was, from an early age, an elitist.

My parents did all in their power to make this so. They sought out "suitable" playmates, pointed out the "nice" girls who had "good" manners, and made me say "Sir" and "Ma'am." After all, they pointed out, we lived on "the right side of the tracks." In a railroad town, that meant something.

By pluck and luck I got into Ivy U., so that I could learn the best that has been thought and said. Somewhere along the line I looked up "elite" in the dictionary. "Elite: a socially superior group; a powerful minority with good taste." Then there was "elitism"—"a powerful leadership or rule by an elite; consciousness of being or belonging to an elite." Doesn't the cream always rise to the top?

Surrounded by other elitists, I burned with a gemlike flame. I greatly admired my academic advisor, an elite guru who passed on the obiter dictum:

> Established in his eminence of taste
> This little man has little time to waste;
> And so the only books at which he looks
> Are books on books, or books on books on books.

One leaves the elitists' camp at great risk. Many will call you a traitor to your class—and treat you accordingly. They will warn you of eternal isolation, ending finally in sterility, disassociation, and mandarinism. On the other hand, you will find that a group of pioneers have over the years dared to leave the beaten path. Van Wyck Brooks made his famous highbrow-lowbrow distinctions in

1915. Shortly afterward, Gilbert Seldes began collecting material on *The Seven Lively Arts* (1924) in order to restore a good conscience to people who enjoyed silent movies and comics, but had an uneasy feeling they should be listening to Puccini and looking at the murals of Puvis de Chavennes.

Other brave souls joined the rebels. Edmund Wilson and Lewis Mumford led the way; John Kouwenhoven and Sigfried Giedion wrote about the unselfconscious efforts of common people to create satisfying patterns, not inspired by ancient tradition, but imposed by the driving energies of an unprecedented social structure.

Despite their impressive work, as well as that of folklorists and anthropologists, most interdisciplinary scholarship that dealt with the popular met with icy disdain. The Big Thaw got under way in the 1960s, when President Kennedy's New Frontier had wide appeal. The heading "Popular Culture" appeared in *Readers' Guide* for the first time in 1960, and the Popular Culture Association was formed a few years later. The goal was to find new apparatus to study and understand the world people inhabit and relish; and to recognize popular culture as a barometer, mirror, and monument.

In her highly acclaimed book *Against Interpretation* (1966), Susan Sontag argued that elite interpretation is the revenge of the intellect upon art. Even more, it is the revenge of the intellect upon the world. She argued for new standards of beauty and style: pluralistic, high-speed, hectic. So did Russel B. Nye in *The Unembarrassed Muse: The Popular Arts in America* (1970). He documented the sharp improvement of popular standards of performance. The simple literalness of Tom Mix and Edward G. Robinson evolved into the symbolic, multileveled popular art of *High Noon* and *Bonnie and Clyde*. To erase the boundaries created by snobbery and cultism, Nye wrote in his final chapter, that have so long divided the arts means, in the long run, greater understanding of them.

Graduate work wasn't designed to erase the boundaries—but to bolster the elitist walls. The higher the holier—the closer you get to your PhD, the deeper the seeds of elitism are planted. One could go and enjoy a "Hollywood flick," or enjoy an outing with "the great American people," but must retreat to the stacks and seminars as quickly as possible. In ivy-covered Gothic buildings, one soaks up the "proper" ideas and phrases. Professors become emperors. Who dares suggest that the emperor wears no clothes?

As a special prize I was allowed to sit at the feet of English professors who taught the New Criticism—an extension of the not-so-new "explication de texte" French school that took words and phrases apart till they squeaked. (This school was later resurrected under yet another name: deconstruction.) We could spend weeks on a single poem—and never ask who wrote it, or when, or where. That's pure elitism.

In due time I found myself wearing the robes of Platonism—that haute couture of the academic establishment. We were encouraged to enter the realm of "value" by looking at ideal embodiments—thus seeking the unblemished display of perfection. We were reminded that that great master of us all, Plato, sought only the True and Beautiful. He abhorred the mob, and labeled democracy the worst form of government. Hence I learned, even as an adolescent, to scorn what was vulgar and low.

Other acceptable labels were aristocratic, purist, highbrow, sophisticate, and (Italian is always in vogue) cognoscenti. I was much given to little magazines and big words. I graduated with honors.

I had little or no sympathy for philosophers such as Parmenides, who wrote: "The time will come when philosophy will not despise even the meanest things, even those of which the mention may provoke a smile." Instead of Parmenides my elitist friends quote José Ortega y Gasset, who thought the masses not only vulgarize and humanize but actually destroy art. They liked Eliot, whose definition of culture was narrow enough to exclude even Pooh Bah; and Ezra Pound, devoted to Chinese ideograms and Mussolini's fascism. Closer to home, Dwight Macdonald thought popular art did not even have the theoretical possibility of being good. He enunciated his Law of Raspberry Jam: Unless checked, the tepid ooze of Midcult will contaminate all civilization.

Then, by a curious set of circumstances, it was by leaving civilization that I began to doubt the wisdom of all the elitists who had indoctrinated me for so many years. It was a pure case of serendipity.

Having a Fulbright grant that took me to Asia, I found myself taking a side trip that fulfilled a childhood fantasy. I landed at the airport in Sarawak, planning to take a trip into the primordial rain forest of Borneo. Roll over, Rover Boys—now it's my turn to deal with the Head Hunters of Borneo! With the black-crested baboon, proboscis monkey, charging rhino, and dwarf buffalo! Which of them might be munching on the exotic rafflesia, whose flowers measure a yard in diameter? And where might the adders, cobras, and scorpions be hiding, waiting to administer their death blows?

I soon learned that the Indonesians have renamed Borneo Kalimantan, and that the airport is more concerned with radar than ritual. But once one goes into the countryside, the history and legends reemerge. There lurks Makassarese, proud old kingdom finally conquered by the Dutch; and the land of the fierce Toraja—never subdued by anyone. And who can conquer the teeming Barito River, flowing majestically into Banjarmasin?

Here enemy heads were hunted and severed for centuries, a fertility ritual to ensure survival. Headhunting was a religious, social, and political requirement. Heads were offered to ancestral spirits at burials and temple consecrations. Sorcerers were sold to other villages to be beheaded. All this is past, my American host tells me. Now Kalimantan is a routine tourist stop, featuring one of the last unspoiled orangutan preserves on earth. He has arranged for me to see it.

That afternoon, I was following a narrow path into the dark jungle. An hour later, we continued to walk, and I continued to sweat. I wiped the moisture from my neck—pleased to note that all the necessary parts were intact. I smiled at the native guides, and told them I felt fine. Inside, I had another feeling. I was scared.

Fear breeds fear. Suppose something *were* to happen—how would we get back? Who could find us? The guides knew the terrain—but how would they cure a scorpion bite? After a sharp turn in the path, we went down to a natural stone tunnel, carved by a tiny stream over the centuries. It was dark, covered with dense vegetation. Down we went, children of the Electric Age, back into the dark cave, the womb. All we could hear, once inside, was the relentless trickle of water.

Finally we emerged into the bright sunlight. As my eyes adjusted, I heard a rustle in the jungle, and saw a dark figure moving toward me. As I clenched my fists, she stepped just ahead of me in the narrow path—a child of six or seven, scantily clad, holding an object in her right hand. She walked boldly straight toward me, thrust forth the object, and spoke: "Coca-Cola, Meester? You wanna cold Coke?"

That was the moment I became an ex-elitist. My Platonic robes fell to the ground, my polysyllabic words left me. I nodded, reached out, grabbed the bottle, gulped. I knew beyond a doubt that it was the Real Thing.

I had not met King Kong but King Coke, and I knew I was about to undertake a new and different quest. The immediate question was: How had she got there, with a cold Coke, just as I came through? What cultural forces were at work, here in one of the remotest spots on earth? How many more people in the world recognize a Coke bottle than an IBM computer or a Patriot missile? Is there any object in the world as universally recognized as the Coke bottle?

I did not see any orangutans, but I shall never forget that adventure. I suffered from culture shock in a distant wilderness. But in a flash I came to realize that America too is a wilderness shared.

* * *

So back I came from the Kalimantan wilderness to my own native American wilderness, where postmodernism has swept the land and we live not in a jungle of wild animals but of electric wires and blackouts. The rate of technological change—and hype—frighten me more than the orangutans. I suddenly want to be on Walden Pond with Henry Thoreau, or on Golden Pond with Henry Fonda. I don't want to be shackled to the computers and satellites which are our new masters. Too late—I write this on a computer, watch the satellite weather reports, and check my e-mail, which has become my new umbilical cord to the outer world. Can I use these devices and not give up all they take away? Is there a livable middle ground?

What I call my terrifying new WWW—World of Wonderful Wizardry—has descended on us like the wolf on the fold. Is there not some parallel, some parable, if we look back a century to another fictional wizardry—and popular frenzy—which L. Frank Baum describes in his 1900 story, *The Wonderful Wizard of Oz*? His book, made into an all-time favorite movie in 1939, features the Wizard, just down the Yellow Brick Road, who could and would cure all problems—courage for the Cowardly Lion, a heart for the Tin Man, a brain for the Scarecrow. Down they all skipped happily with Dorothy (who was dreaming the whole episode)—until they actually got to the Wizard. Alas, he was a fake, a facade, behind which an all-too-human trickster, Professor Marvel, bellowed through a loudspeaker. Dorothy woke up not in the utopian Oz but tornado-torn Kansas.

We are, it seems, going merrily down our Yellow Brick Road. Might the Wizard be Bill Gates? Speaking of failed wizards, how did Adolf Hitler manage to rally Germany behind his Third Reich for twelve years?

An earlier American wizard was Thomas A. Edison, "the Wizard of Menlo Park." He developed many inventions and filed over one thousand patents. He was especially proud of his motion picture machine, which would, he confidently predicted, replace textbooks in our educational system. Hardly.

A lot of wizardry was around in the late nineteenth and early twentieth centuries. The Three Ts (telegraph, telephone, typewriter) set the stage. Then in quick order came the Kodak box camera (1880), gramophone (1888), radio (1895), diesel engine (1897), and Ford's mass-produced car (1908). Each promised to change our way of life, and indeed did so. Educators in particular heralded each new utopia. The radio, we were promised, would replace the blackboard. So would the reel-to-reel tape recorder, film strips, videotapes, and overhead transparencies. Now it's the computer's turn.

"Computers in classrooms are the filmstrips of the 1990s," Clifford Stoll writes in *Silicon Snake Oil: Second Thoughts on the Information Highway* (1995), recalling his 1960s school days. We loved them, Stoll says, because we didn't have to think for an hour, and teachers loved them because they didn't have to teach.

Is there a pattern here? Can the fate of some of these early wonders (now collecting dust in attics around the country) which looked less wonderful when the next wonder occurred, teach us a lesson?

Clifford Stoll, Neil Postman, Larry Cuban (who wrote *Teachers and Machines: The Classroom Use of Technology Since 1920,* 1985*),* and Todd Oppenheimer join a growing number of educators and critics who do see a pattern. Their analyses vary, but all outline a similar progression:

1. The dramatic announcement by the promoters, backed by big promises and big bucks. (Bill Gates is said to have spent $6 million to launch his Windows 95 campaign).
2. Pushed by public opinion and school officials, teachers quickly adopt the magic package, but find it does little or nothing to improve academic achievement.
3. Other factors—not the new item—get the blame: incompetent teachers, insufficient funding, students with home problems, unwieldy bureaucracies, etc.
4. Finally everyone has to admit the new technology doesn't do the job.
5. Lo and behold, just in the nick of time a new solution appears. The next generation of technology repeats the same old platitudes and the whole cycle starts again.

But *this* time is different, proponents always say. There has been evidence of more enthusiasm—even euphoria—about computers on the part of administrators, teachers, and students. The fact that in many instances the new equipment has been donated by corporations or funded by major federal and state grants is a factor. So is the fact that a blue-ribbon panel appointed by President Clinton (who, one hears, can't even type) are extravagant in their praise and projections. Even the name is, in our jargon-filled bureaucracy, impressive: the National Information Infrastructure Advisory Council. It took thirty-six members to staff a committee with that title and all were unequivocal computer boosters. Two-thirds of them worked in the high-tech and entertainment industries. Skeptics were notably absent.

Was there any doubt what they would recommend? And could anyone be happier than Bill Clinton and Al Gore, who intend to use the computer to help keep them in the White House? Predictably, the demise of the blackboard is again announced. President Clinton himself has promised to build a bridge to the twenty-first century, where computers will be as much a part of the classroom as blackboards. His program would, according to estimates, cost between $40 billion and $100 billion over the next five years. By then, we are told by the New Utopians, we shall all be linked in the global village. What does this mean?

These claims should be put in a broader perspective—one that goes far beyond five, fifty, or five hundred years, beginning when Homo sapiens left the cave and started on the long journey to cyberspace. What approach will serve me best? History, anthropology, folklore? They are all new, if we consider how much occurred before these "disciplines" came into being. Why not go back before writing, science, print, even before the alphabet—how can this be done?

With popular, or common, culture, where elitists seldom look. Cicero said we are all linked by a common chain—workers in Sumer to the Silicon Valley; wives and mothers in every clime, invaders from Xerxes to Saddam Hussein. Everyone everywhere has birth-life-and-death problems. Popular culture contains all these. But just what is "popular culture"? How to approach it? What terms, themes, links will emerge? Perhaps, as an ex-elitist, I can make a fresh start.

It should be quite a journey.

Words, Words, Words

In the beginning was the word.

John 1:1

Words are signs of natural facts. But what is a word? A fact? And when is it "natural"?

Words, Samuel Johnson wrote, are the daughters of earth. We earthlings are both the prisoners and promoters of words—spoken, written, printed, carved, glorified, electrified. Words make us human. History was still blind when words filled the air.

How to explain their origin? Five popular theories offer possible answers:

1. The bow-wow theory. Words originated through imitating natural sounds (onomatopoeia).
2. The poo-poo theory. Speech originates from involuntary expression of emotions.
3. The yuk-yuk theory. Words arose from the chance sounds that are associated with special and important events.
4. The sing-song theory. Words grew out of primitive and wordless chants.
5. The yo-heave-ho theory. Words developed from grunts and sounds of physical exertion.

We shall never know which, if any, of these theories are correct. (Or, indeed, if all of them played a part.) What we do know is there are more than 3,000 spoken languages extant today, with who knows how many millions of words. Words have no static meaning—in fact, they never mean exactly the same thing twice. The time, place, context, inflection, stress, and audience can't be duplicated. The meaning of the word is what the speaker intends to be understood by listeners.

Yet that sound can be a trumpet call. Words are priceless tools and motivators in the hands of master orators. How we all would like to have heard Demosthenes address the Greeks, or Cicero the Romans! But at least we can, thanks to our technology, still hear Roosevelt speak to the Americans, Churchill to the British, Hitler to the Germans. Their words are magic. They can send people out to do deeds that are beyond belief.

Words can also put people to sleep. As a professor I know this all too well.

I am a wordmonger by birth, a poet by preference, and a teacher by profession. Fishmongers catch fish, ironmongers reshape iron, wordmongers work with words. In the *Oxford English Dictionary* there are whole columns on "word," which has been spelled wurd, wored, woerd, wuord, wort, werd, wurde, wird, and countless other variations. As for the meaning of "word": Any or the least utterance, statement, or fragment of speech; anything at all, said or written. I make my life, and my livelihood, by speaking and writing words—and am proud to be called a wordmonger. So have been many others before me.

In *Hot and Cold Bath* (1709), Sir John Floyers wanted to find "some strong young mongers of good constitution, who have brushed through misfortunes." Robert Southey sought out other "humanity mongers," who would oppose capital punishment. In America, introducing his *Twice-Told Tales* (1837), Nathaniel Hawthorne wrote, "I must be content to bear the stigma of a fiction monger." One of the greatest prose writers of our time, Winston Churchill, liked the sound of "wordmonger." I am pleased to share that sentiment.

Words make us human. Wordmongering is the most complex and crucial thing we do, whatever our state or status. When you're out of words, you're out of life. In primitive societies words are magic—physical instruments that cause things to happen. Magic is as natural as breathing; bone-deep, all-pervasive. Words are instruments of power, and oracles depend on them.

Written words substitute for sound. Their great advantage is that they last long after the sound dies. An eighth-century Chinese poem observed:

> That art is best which to the soul gives no bound,
> Something beyond the form, beyond the sound . . . .

Electronic words, on screens and laptops, vanish into thin air; when printed, or put into books or archives, they endure. Print is the medium of history. Radio may be our ears, television and computers our eyes, but print is still our memory. When we say we have "been in touch" with someone are we only saying that the echo chamber has worked? That we have heard coming back to ourselves the gratifying echoes of other voices? In the Information Age, are we overwhelmed with words that come and go with the speed of lightning? Have we lost some of the great advantages of the Print Era?

Print began slowly over 5,000 years ago with the development of pictographic, ideographic, and hieroglyphic symbols. Then came phonetic writing and the formation of alphabets. But the great breakthrough—many would say the beginning of modernity—came when a German named Gensfleisch, using the name Gutenberg, invented movable type in the 1440s.

The basic experience of humanity has been reshaped by the invention of type. We left behind the totally oral phase of human history, of preliterate tribalism; and the age of script, begun in the fifth century B.C. by the Greeks. Print tended to shut the door on oral (ear culture) and empower literate (eye culture), thus changing our sense ratio, perhaps even our understanding of the sacred and secular. It provided the first assemby-line, mass product. Type was the prototype of most of our later machines. The more rapid and extensive spread of information made possible the Renaissance, Scientific Revolution, and Reformation. It also encouraged capitalism, communism, nationalism, and individualism. Music went from Gregorian to linear. Life was no longer a collection of separate segments, but (like print) a continuity. When Martin Luther's German translation of the Bible was put into the people's hands, Western religion would change forever.

Books are our communal memory and nourishment for our souls. They are, at the same time, the building blocks of our popular culture.

Words, like bombs, can explode. They can lie dormant for years, then become buzzwords—on everyone's lips. One prime example:

virtual. For years it slept quietly in the dictionary, having virtually no users. "Virtual" meant "being so in effect, although not in actual fact or name." The adverb "virtually" meant "in effect although not in fact." Virtual contrasted with reality. Now it is a link word: virtual reality. And we have a paradox: how can virtual be real?

Media fantasy has become so realistic, and real technology so fantastic, that what used to be clear dividing lines have either blurred or disappeared. Things are seldom what they seem.

From the standpoint of the life-world, some critics argue, the technological age is an age of death—soon we will have no history, politics, or economy outside of our video terminals. The flash which places us at the center of the screen, John O'Neil writes in *Plato's Cave* (1991), renders us the outcasts of mankind's hitherto civilized incorporation and continuity.

Words, then, are both our target and our ammunition. Putting my thoughts (and those of many others) into readable words is at the core of our enterprise. Words are our great treasures, and should be used accordingly. I shall try to use them judiciously, so that I may hit the target; to say what I mean, and to mean what I say.

The Prince of Pop

At some point in the early morning of the last day of 1980, Marshall McLuhan died peacefully in his bed in Toronto, Canada.

He was a young sixty-nine—a brilliant, baffling man who had mesmerized and electrified a whole generation, then fallen out of favor and joined the ranks of fallen stars.

I was one of the many pilgrims who made their way to visit the Canadian guru who claimed to have no point of view, but to make probes; to talk back to media and set off an adventure of exploration. What an adventure it was!

Who could forget that office—a former coach house out back at Toronto University, with a picture of the Marx Brothers, his Cambridge rowing paddles, and this sign on the wall:

> INFORM ALL THE TROOPS THE COMMUNICATION
> HAS COMPLETELY BROKEN DOWN!

We had to wait eighteen years after his death to get the first accurate and responsible narrative of McLuhan's life—Philip Marchand's *Marshall McLuhan: The Medium and the Messenger* (1998). Now we have a responsible guide through the weird, exhilarating vortex of ideas that Marshall McLuhan gave us.

Born on the prairies of western Canada in 1911, Marshall McLuhan lived in seven houses in the first eight years of his life. He later spoke of the "mean mechanical joyless rootless" life he grew up with. His father sold insurance, but his ambitious mother (who appeared on the stage as an impersonator) saw to it that young Marshall had a brighter future. She sent him off to the University of Manitoba (1928-1934) and then to England's Cambridge University (1934-1936). There he studied with I. A. Richards, the uncrowned king of English studies, wrestled with the contorted prose of James Joyce, rowed on the college team, and returned to the United States after getting his MA to take a teaching assistantship at the University of Wisconsin.

He was, in those early days, no friend of pop culture. He found most of it monstrous and sickening and said that he studied it for his own survival. Yet, in due time, it would ensure not only his survival but meteoric fame. McLuhan changed the whole focus of media studies and popular culture.

Fascinated with electronic media, which compressed space and time into space-time, he began to think of a world soul. Perhaps this mystical secular image came from his conversion to Roman Catholicism in his early twenties, and from swallowing the ideas of G. K. Chesterton whole. If print had divided us (into interest groups, castes, and nations), perhaps electronic magic would unify us. Instead of living in warring nations, we might live in a peaceful global village. In McLuhan's words, we might translate our lives into a "spiritual form of information," bringing the entire human family into a single consciousness.

The young Canadian became fascinated with television. It did not please but irritated him, yet entranced his students. The works of a Canadian scholar, Harold Innis, opened up new vistas. Media, in whatever form, had always been a key to dominance. Rather than scorning the new technology, he must study and demystify it. He began to formulate his most famous idea: the medium is the message.

Advertising personnel, the television industry, and academia made him a celebrity, a guru. Disc jockeys got into the act, shouting, "What are ya doin', Marshall McLuhan?" He even made a cameo appearance in a Woody Allen movie. Gradually his star shot across the celebrity sky and began to fade. Having become popular, he was plagued by the problems that always plague such people—they fall as fast as they rise.

After 1968, a number of critics and communicators believed that despite his sudden fame and peppery puns, the Canadian Guru had misread history and even misunderstood the medium where he seemed most brilliant—television.

They insisted that the role of content to television was crucial, just as it had been with the printing press and every other great innovation. McLuhan's ideas, some critics insisted, were based on wild speculation, not observation.

Richard Schickel led the pack with a highly critical 1968 article ("Misunderstanding McLuhan") and Everette Dennis published his "Post-Mortem on McLuhan" in April 1974. James Grunig edited a volume called *Decline of the Global Village* (1976), attacking McLuhan's "confusing and contradictory style of writing." Others pointed out the "logical inconsistencies." Some of us who owed much to McLuhan deserted him. Alas, I was one of these, publishing a chapter called "Goodbye, Guru" in my *Parameters of Popular Culture* (1978).

McLuhan wrote me with a reminder and a prophecy. "Have you forgotten that we live in an age of nostalgia and that all past fads have to be retrieved frequently?" He was right. The MIT Press republished his 1964 classic (*Understanding Media: The Extensions of Man*) in 1994. In his perceptive introduction, Lewis H. Lapham points out that much of McLuhan makes a good deal more sense in 1994 than it did in 1964. He forecast the coming and meaning of MTV, the Internet, shopping malls, and e-mail before they existed. He also forecast that the "global village" which he had anticipated and named would either lead to a new interlocking harmony or a worldwide bloodbath. He seems to have been right on both scores.

McLuhan's great contribution was in foreseeing cultural upheavals. However outrageous or irresponsible his books may have seemed three decades ago, they were remarkably accurate—we live in a McLuhanesque world. He ranks with cultural prophets such as Henry Adams and Aldous Huxley, whose 1932 *Brave New World* looks more and more like our world. And what about George Orwell's *1984*? Will Bill Gates be our Big Brother?

Important new material on McLuhan is emerging. In addition to Marchand's book, McLuhan's daughter Stephanie has collected six hours of videotape, with commentary by Tom Wolfe. Publications such as *Wired* and *Whole Earth Review* have changed McLuhan's "spiritual form" to "cyberspace," added some jargon of their own, and taken off—repackaging the man many have been discrediting for decades. Did Wolfe ask the right question? Suppose McLuhan *is* what he sounds like?

ENDNOTE

Marshall McLuhan's primer of popular culture assessment is *The Mechanical Bride: Folklore of Industrial Man* (New York, The Vanguard Press, 1951).

Dozens of magazine and newspaper articles on McLuhan appeared in the 1960s, but it was not until 1998 that we got a full-scale narrative of his life—Philip Marchand's *Marshall McLuhan: The Medium and the Messenger* (Cambridge, MA, MIT Press). McLuhan's daughter Stephanie is issuing tapes of her father's appearances and interviews, and interest in him is increasing. Consult the *Readers' Guide* for recent articles.

Anti-Pop

We come neither to praise nor to bury popular culture—instead, to try to understand its history, function, and role. The critics have strong cases; they must have their day in court. Every age has echoed Plato's fears that the rule of the mob is the worst possible form of government; that the ignorant will choose to spend their lives gazing at unreality in the Cave. Our Cave might be the television or computer screen.

Since the big fish swallow the little fish by supreme natural right, we can always expect strong protective measures and actions by big elite fish. Popular culture alarms them. In our polarized electronic society, antipopular culture forces have multiplied, and the alarm has been sounded. From the fundamentalist right and the liberal left, the criticism has increased. With the new means of disseminating information, it greatly multiplied. This means the medium and money are the message.

American culture, some insist, has become a crowd culture, dangerously trivial. The people become the crowd or the customers. Their sole function is to consume. We must examine these charges.

One of the most interesting critiques comes from the "decline and fall" school. Once, pop culture aspired upward—now it's all downward. One of the most persuasive critics of this school is Richard Corliss, who elaborated on his position in the June 8, 1998 issue of *Time* magazine. Corliss divides our century in half. The first half had a vibrant, energetic popular culture that imitated the upper class; the second half, the underclass. As the homogenous culture evaporated, the divisive underculture took over.

Pioneering two new art forms, movies and popular songs, "first half" people concentrated on love, success, assimilation. They aimed for universal acceptance in a homogenous culture, wanting everyone to sing their song, see their movie, try their dance. The popular thirty-two-bar easy-to-hum song by someone such as Irving

Berlin or Jerome Kern gave you a lift. June, moon, croon, and spoon were cozy partners. Frank Sinatra, Rosemary Clooney, or Ella Fitzgerald sang, while Gene Kelly, Fred Astaire, and Ginger Rogers danced their way into our hearts.

There were really funny movies—with the Marx Brothers, Buster Keaton, W. C. Fields—and truly epic movies: *The Ten Commandments, Ben Hur, Gone with the Wind.* Cowboys, the guys with the white hats, took care of the bad guys in the black hats and preserved the honor of visiting school teachers from Boston. We had a culture we could believe in and count on.

Then came the deluge. A convenient date to divide the century is 1951, when Marlon Brando brought Stanley Kowalski to the screen. The beast was unleashed—the Wild One was ready to take the road, rockin' and rollin' along the way. We moved from grace to grunts, manners to menace, tux to torn T-shirt. Now there were two poles—Fred Astaire and Marlon Brando. Astaire gave us class, Brando gave us sex.

Nothing changed more than pop music, which the young claimed as their kingdom. Smooth singing (with Sinatra) gave way to snarling (Bob Dylan); elegant dancing (Ginger Rogers) to wiggling (Tina Turner). Parents no longer understood their children—and vice versa. Come home at midnight? Make me! Moral standards disappeared, as did homogenous culture. Diversity brought back segregation and gender obsession and led to the Age of Victimization. For better or worse, the generation gap looms large in today's thinking and acting.

Pursuing the popular reminds us of poor Sisyphus, trying to get the stone to the top of the hill, always failing. Once we think we have solved the puzzle of popularity, having got the stone to the hilltop, it rolls back down to the bottom. Winner lose all.

Or—to use another analogy—the puzzle has many pieces, like an intricate jigsaw game. Some pieces are always missing, and others don't fit. Essential questions remain unanswered. How did we get from the Virgin Land to the Raped Environment? From the Deerslayer to the Cop Killer? From the Land of the Free to the Land of the Peeved?

How did Elvis become the King? Or Marilyn Monroe the Mona Lisa of the movies? How did Andy Warhol become the best-known artist in the world? Why have so many Americans given up on democracy?

Not only celebrities and fads but whole technologies get swallowed up in the Black Hole. The Golden Age of radio lasted only twenty years, and the Golden Age of Hollywood passed even quicker. Look in the attic and see how much of today's new technology is tomorrow's junk: the reel-to-reel tape player, black and white TV, 8 mm film camera and projector, 78 rpm records, both manual and electric typewriters, etc., etc.

Life in the twentieth century has put us on a roller coaster: ups and downs to take away our breath, our balance—even our lives. No century (or so we think) has been bolder, bloodier, brasher. Everything nailed down is coming loose. The earth changes as we walk on it.

The age of instant issues . . . endless crusades . . . posters and protesters unlimited. We have been for and against so many things that we aren't sure where the action is, what the issues are, or who's on whose side.

At the heart of popular culture are myths and religion. No one can move the people like a good preacher. George Whitefield, the Methodist itinerant, was said to bring a vast audience to tears, merely by intoning the word "Mesopotamia." Others played a rougher game that resulted in "barking, jerking, and other primitive traits," using what came to be called the "Baptist whine." Popular tunes were turned into evangelical hymns. Emotion, free expression, and music were a mighty trinity. They still are. Go to a rock concert and see for yourself.

To identify with any new phenomenon—such as a rock concert—you must work at it. Culture isn't inherited, but learned. Communities are complex things, with their own structure, codes, goals, traditions, and memories. When you change cultures you become not a participant but a spectator—often a baffled one. Just watch a foreign visitor trying to understand an American baseball game. Then go to England and try to make sense out of a cricket match.

Without knowing it we lived though a great revolution in 1989 when Communism collapsed. Old taboos were junked, old clichés abandoned, old tyrants rebuked—with very little bloodshed. What made this revolution unique is that we all saw it—on television.

Spies came in from the cold, people went out into the streets, and the world watched long-stable systems topple. What did it all mean?

Revolution is popular culture going at top speed—the "voice of the people" shouting. When that happens, watch out. Today's revolutionary mounts the guillotine tomorrow. Long live the revolution.

Those of us who were print-trained must realize that a new generation has been trained in hi-fi sound and cinematic vision. Their ears expect a tonal perfection no live performance can match. Their eyes respond to twenty-four frames a second; movies and TV foreshorten time and space and flatten reality. For them all the world is not a stage or a textbook; it's a film clip or a sound bite. They will never understand how we think and react. The past overrides them on the saddleback of the supersonic present.

For them our maps are meaningless. Their world doesn't fit maps of our making. Why? Because they are no longer earthbound. They think like astronauts, and have no idea why Greenwich should keep the time. This changes everything. *How* we perceive creates *what* we perceive.

For many young people the world not only looks but sounds, smells, tastes, and feels about the same everywhere. Theirs is a culture of homogenized sameness. They are what they buy and the buying never ceases. Think of Sisyphus with a credit card, buying a burger and a Coke in Outer Mongolia.

The seasons, which used to influence everything, have virtually been reduced to a mere change of scenery, as have night and day. We can get all food in all seasons at any hour and be in any other climate before the sun sets. Suburban and sub-rural blur; basically all brands are the same brand with a different wrapping. This is the world of Mickey Mouse, Michael Jordan, and Madonna.

Fame and the Famous

All is ephemeral—fame and the famous as well.

Marcus Aurelius,
Meditations

Marcus Aurelius may never have heard of popular culture but he summed up its chief dilemma in a few words. So did John Milton, who said that fame was a spur. Though we have traded horses for space shuttles, our riders still feel that spur.

In the past, only a few great leaders, warriors, and saints were famous. Now the term is used indiscriminately for all manner of men and women. Recall Andy Warhol's prediction: One day we'll all be famous for fifteen minutes. We will be like shooting stars in the heavens. In fact, we often call famous people "stars."

Why do we yearn to see stars, on earth as in heaven? Why do we spend time and money to glimpse a Hollywood star, scream at a rock musician, vote for a political winner, hum the Number One Hit Tune? Do we lurch for the glimpse because we know popularity won't last any longer than water captured in a cupped hand? We know it will slip through our fingers, and soon be gone—no matter how thirsty we may be. But our thirst persists.

We began the wishing and wondering in the nursery: "Twinkle twinkle little star, how I wonder what you are . . ." We soon learned of the Three Wise Men, who followed their star to Bethlehem. Later on, we were told that Caesar was as constant as the northern star; that Romeo and Juliet were star-crossed lovers. Milton greeted "The bright morning star, day's harbinger," and Emerson urged us to hitch our wagon to a star. We got a vision of how exciting that might be when we first saw van Gogh's *Starry Night.*

We never know when and where the next star will appear; but we can recall several from the recent past. Perhaps they will help us in the task we are undertaking.

* * *

On the morning of October 19, 1987, an eighteen-month-old baby named Jessica McClure in Midland, Texas, became famous. Her picture and her story were flashed around the world. Her mother, Reba, had left her unattended, and Jessica had fallen down an abandoned well. Now, sixty hours later, she had been rescued. On hundreds of front pages little Jessica, just up from the well, was seen in Reba's arms, while Chip (Jessica's father) seemed as useless as Joseph in the Christmas crèche scenes of the infant Jesus. But we were in Midland, not Bethlehem—and the world was watching, awestruck.

Then–Vice President George Bush and his wife Barbara rushed to the scene. They were photographed in Midland Memorial Hospital, gazing intently at Jessica. Shutters clicked, cameras rolled, Bush spoke to the reporters: "This is what America is. The rock was hard, but they heard the voice and they didn't give up and for sixty hours they dug."

Of course we knew all this. Ever since Jessica fell into the well the world had been put on hold. When she was hoisted out of the shaft, at 8 p.m., all three major networks interrupted their prime-time schedules to show the final minutes of the rescue. Nations around the world joined in the wild acclamation. Reporters and cameramen in Midland far outnumbered rescuers. Then Mother Reba spoke: "Thank you, America." She carried the blessed babe back to their small frame house, now bedecked with a huge banner: "WELCOME HOME JESSICA. WE LOVE U."

The congregation of the Cotton Baptist Church stormed the house en masse. Local florists ran out of Jessica wreaths. The hospital where Jessica was resting was inundated with balloons, money trees, prayer rugs, and stuffed animals. Employees of Disney Studios rushed a giant Winnie the Pooh bear to the frantic scene.

All along the way, Federal Express employees inscribed loving greetings to Jessica. President Reagan made a personal call to Reba and Chip, assuring them that the White House rejoiced. Reba was inspired to make another statement: "We're going to have a big bash for this little girl."

A tough Texas roustabout, covered with Texas crude, appeared on national television to admit he had cried. Other enterprising

Midlanders were too busy to cry. Business boomed, sales soared. A special crew was rushed to the hospital to handle the incoming teddy bears. The post office sought guidance on how to handle mountains of incoming mail. Texans gave the postmaster a good piece of Texas advice: a man must do what a man must do.

Then suddenly the show was over. One by one the cameras and the reporters disappeared. A lingering newsman spotted Tim Parker, one of little Jessica's rescuers, wandering about the site—bare ruined rigs, where late the cameras rolled. Walking through a pile of rubble, Tim found one of the drill bits used in the digging.

"I guess I'm a hero," he said. "I guess we're all heroes."

The reporter asked him what he planned to do with the broken bit. "I'll give it to Reba and Chip," Tim said, "with a note saying, 'For Jessica—sixty hours of love.' "

One last shot finished the Midland saga. It was of Jessica's grandfather, whose leathery face had been baked by the relentless Texas sun. During all these earth-shaking hours, he had never left his house to watch the rescue.

"I stayed home and watched it on TV," the old man said. "You can see more that way."

* * *

Popularity comes and goes like a meteor in the sky, a thief in the night. It affects almost everything: even the toys our kids play with. Do you remember when G. I. Joe dolls took over toydom, only to be pushed aside by the Cabbage Patch Kids, then *Ghostbusters?* And what about Teenage Mutant Ninja Turtles, popping up out of sewers in the 1990s? Would they be the most popular cultural fad since "Beatlemania" in the 1960s? During March 1990, reported *People* magazine, three of the ten best-selling children's videos were "Turtle" texts: *Killer Pizzas* number 4, *Hot Rodding* number 8, and *Heroes* number 9.

Turtlemania! Who could have predicted, produced, or explained it? Of all the creatures that might invade the American psyche, why was the answer "turtles"? Yet they kept coming, and the public kept buying. Moving out of shops and stores, they invaded our television and film. *Teenage Mutant Ninja Turtles: The Movie* became the record offering for the 1990 summer season. Stores couldn't restock their shelves quickly enough. Sociologists tried to explain what was happening. Characteristically, they failed.

The hunger for all things turtle seemed insatiable—food, clothing, jewelry, ornaments, books, toys, souvenirs. Scholars of popular culture, often slow on the uptake, finally saw the light, and translated the obvious into the academic: "Here is the new device for transmitting messages about the dynamics of the postmodern world." Mercifully, they didn't attempt to dissect or to define a postmodern turtle.

Then the search for earlier prototypes and models began. Were the postmodern turtles so different from the less modern cats, wolves, and roadrunners that had captivated children's lives in earlier times? Was this a recycle? Had wolves abandoned sheepskin and put on turtle shells? And who could explain the "Mutant Ninja" part? To all this the newly rich turtle peddlers may well have replied, "Who cares?"

"Turtle" sales reached $400 million in 1990—then declined 40 percent in 1991. One retailer, looking at huge stacks of unsold and unsalable turtles, gave the ultimate explanation: "Kids finally got turtled out."

Fads and fancies dance to their own rhythm. They can no more be explained than sudden changes in the weather or clothing styles and pop tunes. The more we study popular culture the more baffling and unpredictable it becomes. Perhaps that helps to explain its perennial appeal.

While the turtles were invading America, Saddam was invading Kuwait. The whole matter came up as quickly as a sandstorm in Arabia—and threatened to cut off our supply of Kuwaiti and Arabian oil. That, of course, meant war.

War? What we all feared was another Vietnam. President Bush assured us we would not have one, but we had read his lips on other issues, and weren't convinced. And what about the deep-rooted antiwar sentiment, the burned American flags, the Peace Movement? Who would fight and win a war in the distant desert? Who could lead American troops in so dangerous a task?

Suddenly there was an answer: a huge hulk of a general with a ruddy face, giant hands, and a name befitting a German field marshal: General H. Norman Schwarzkopf. Where had he been all our lives? Never mind. He was our new shooting star, and we knew instantly just what his nickname should be: "Stormin' Norman."

He stormed. First he assured us that we had the right stuff; that Iraq's Saddam Hussein had the wrong stuff; and that he would set a timetable and win the war in double time. He taught us to recognize the Apaches, Warthogs, and Cobras that would win the war, and assured us that the KIAs (Killed in Action) would be extremely low in the KTO (Kuwait Theater of Operations). We were more than impressed: we were patriotic. What began as Desert Shield would soon become Desert Storm. And in the center of the stormin' would be our Norman.

The press called him the Unknown Soldier and set out to find out all about him. He had a good war record in Vietnam, but little name recognition. In 1990, he got that in a hurry. Finally we had a leader who was blunt, not glib; passionate, not packaged; a real John Wayne leading a real, tough army. People started making comparisons. He reminded us of the "rough and ready" generals before the Civil War; or of Teddy Roosevelt, who led his victorious Rough Riders across Cuba. "The general is commanding, quotable, and has a great personality," Ronald Reagan's former speechwriter, Ken Khachigian, observed. "He's got all the makings of a popular hero."

There was more. He had a cheerful wife named Brenda, three happy, healthy children, and a black Lab named Bear. *Jeopardy* and *Cheers* were the general's favorite TV programs, Breyers mint chocolate chip his favorite ice cream. He could be anything to anybody. He was the right man at the right place at the right time. Who could ask for anything more?

After months of waiting, shipping supplies, and watching we were ready to confront the enemy. Saddam wouldn't budge, and promised us "the mother of all battles." We were also playing Russian roulette—what would the Soviets do if and when the going got rough? General Schwarzkopf made it clear what we should and would do: Go for it!

A hundred hours—not days—after our full-scale invasion of Kuwait began, the war was over. The general had delivered on his promise—we had won in less time than it took to fight many of the battles in World War I or II, and with fewer casualties than we suffered on America's highways during the Persian Gulf War.

All this produced the greatest outbreak of "flag-itis" in American history. The number of flags surpassed visual comprehension. At Arlington National Cemetery, for example, 210,000 were distributed in preparation for Memorial Day services on May 27, 1991.

By then General Schwarzkopf had come home, landing at Tampa's MacDill Air Force Base. Hysterical fans wore "STORMIN' NORMAN FOR PRESIDENT" T-shirts—although no one knew for sure if he was a Democrat or a Republican.

In any case, he was a winner, a man's man, America personified. His star was at its zenith. New York City gave him the mother of all ticker tape parades to confirm it.

After that the general, whose military strategy was flawless, made the serious mistake of overexposure. He burned up his patriotic fuel and moved ever closer to the Boring Zone. One of the first to perceive this was Amitai Etzioni, a professor at George Washington University. In the end, he said, we can't stomach a real hero. We're happy to have heroes for a week, but we have to tear them down.

We began to see the Conqueror in a different light. He posed with Mickey Mouse in Tampa, and the Ziegfeld Girls in New York. Would he go on to endorse Excedrin, Jell-O Pudding Pops, and Bud Light? Was it true that Bantam Books had offered him $5 million for his autobiography? Would he give a carefully regulated number of speeches—for $50,000 each? Was General Schwarzkopf going to be the victim of General Marketing?

The answer turned out to be no. He did not seem anxious to expand his power position like so many third-world soldiers around the world, or repeat General Douglas MacArthur's mistake of outliving his usefulness. He did not pretend to have solutions to a host of America's pressing internal problems. He was not dazzled by his own four stars.

In August 1991, after thirty-nine years of service to the U.S. Army, H. Norman Schwarzkopf retired. It showed he knew something about shooting stars, as well as shooting guns. He left the scene with grace and style. Like our greatest soldier, General George Washington, he retired before others suggested that he go.

That cleared the heavens for the next famous star, and added a rich chapter to American history and popular culture.

As different as Jessica, the Teenage Mutant Ninja Turtles, and the general are, they have this in common: each was a star in the popular culture heaven. Can we explain their common denominator, and the process that gave them each a brief, bright moment of fame?

There is no easy explanation, but there are clues. Each came at the right moment—when people were prepared for them. Each was given extraordinary and continuous attention by the media. Each became, for millions of people, worthy of attention.

How did the process begin? The consensus among social scientists (for whatever that is worth) is that the process begins with the media. If they can't tell us *how* to think, they can and do tell us over and over again *what* to think. To survive and attract the readers, clients, viewers, or advertisers needed, they must always pose this question: *What is the most important issue or problem facing America today?*

Generally it turns out to be one with which few Americans have had firsthand experience: hostage-taking, the Mafia, S&L bailouts, terrorism, or genocide, for example. If none of this magnitude is available, then a lesser and more manageable one must be served up—such as Jessica or the Turtles. Once the "issue" or person is in place, we call in the pollsters, who always confirm the prearranged message. Yes, this *is* "our most pressing issue." Prove it by polling! Polls have mushroomed in recent years; now the networks and most major newspapers have their own polling services. In this way public opinion is merely turned into statistics and fed back to the public.

Then the handlers, hucksters, talk show hosts, and "experts" can take over. Their voices, and not the voice of the people, are heard throughout the land. Old-fashioned debate, dissent, and analysis are pushed off the stage. People who know nothing about the issue are asked for their opinions, and ignorance is shared nationwide. The poorly informed drown out those who might actually know. Collective insecurity joins hands with competitive hysteria, and the cycle takes over: from alarm to concern to frenzy to boredom. We can't stand that—so the whole process starts again.

Shooting stars, as we have defined them, come not only in patterns but cycles. This may be the hidden fuel that propels popular culture. But before we say any more about the propellant, perhaps we should try to define the thing propelled. That will be no easy task. We may have to call on Alice in Wonderland for help.

ENDNOTE

The *Journal of Popular Culture,* now in its thirty-second year, has been wrestling with the definitions and function of popular culture since its 1967 inception. The Popular Press, also publishing from Bowling Green State University, has produced scores of books dealing with all aspects of this field. One of those books is my *Parameters of Popular Culture* (1978), in which my own thoughts are given more extensive treatment. See also my *Go, and Catch a Falling Star: Refiguring Popular Culture* (New York, American Heritage, 1994). Ray Browne's *The Hero in Transition* (Bowling Green, OH, Popular Press, 1995) presents a collection of essays on the subject.

The Mosaic As Metaphor

The term "mosaic" is the word for both an ancient art form and potential modern metaphor. Derived from the Greek *Mouseios*, "of the Muses," it has come to mean a picture or design "made in mosaic," and can apply to literary compositions. All mosaic makers are trained to express and compress. They capture, reflect, absorb, splinter all the light, opening up multiple meanings. Surface fracture is crucial. Light is multiplied, split into an infinite number of chromatic units, as with prisms.

The tiny pieces, called tesserae, in great mosaics were made of enamel, glass, marble, precious stones, mother of pearl—each with a unique shape and reflection. Three essential ingredients—light, rhythm, and color—were skillfully blended so that no one dominated. Colors became more than mere colors: sparkling gold, shimmering silver, pulsating blood-red, emerald green, murmurs of milky white. Awed and amazed, poets summed it up as best they could:

> *Aut lux nata est aut capta. Hic libera regnet.*
> *(Either light was born or imprisoned here. It reigns supreme).*

I think Europe's greatest mosaics are in Ravenna, Italy. No wonder Dante wrote his *Divine Comedy* there. Dante's literary mosaic, moving from earth through hell, purgatory, and paradise, has awed and inspired readers for centuries. Like the stone mosaics in Ravenna, which he knew well, Dante transformed space, dematerialized structure, and produced his masterpiece, *The Divine Comedy.*

Others have tried to make literary mosaics, with mixed results. My own immediate model, Marshall McLuhan, made his mosaic of the modern media, and became one of the most controversial figures of the mid- and late twentieth century. He in turn was inspired by two mentors, James Joyce, whose *Ulysses* (1922) is our ultimate mosaic, and a Canadian academic, Harold Innis, who set out to arrange his new insights into mosaics that reflect our own time.

Innis presents his insights in a mosaic structure made up of seemingly unrelated and even disconnected sentences and aphorisms. His new patterns are not made for the traditional linear reader. They demand an acrobatic wit and agility. Innis's most influential book, *The Bias of Communication* (1961), is already a modern classic.

Searching for a way to encompass centuries of popular culture, and to allow the freedom to move quickly from era to era, subject to subject, I decided to follow the lead of these mosaic makers. A "mosaic structure" leading to patterns seems to me well suited to our age and my subject. In mosaics, separate pieces create a complex multilevel awareness. The glory of past ages, they are crucial to our electronic perceptions. The television screen and computer can be seen and studied as our latest mosaics.

Certain insects have what biologists call "mosaic vision." They have compound eyes made up of independent units, so that objects viewed are not seen in isolation but in patterns.

We also have today "aerial mosaics," involving photographs that are matched so as to show a continuous photographic representation of an area. Can we create an aerial mosaic of popular culture?

Is the mosaic metaphor being widely used—in what context? I checked on our library computer system and gasp—there are 110 books with "Mosaic" in the title! The word has a new popularity of its own. Here are some examples.

There are mosaics about places (Mexican Mosaic, Taos Mosaic, Urban Mosaic, Balkan Mosaic, Mountain Mosaic), periods (Renaissance Mosaic, Elizabethan Mosaic, Preindustrial Mosaic, New Deal Mosaic), and medical mosaics about various viruses, cell types, and diseases. Mosaics on Mohammed's people, Jewish perceptions, and Mormon settlements—on and on we go.

"Mosaic" is becoming a companion term for the buzzword "diversity." The once-popular melting pot metaphor is out. Salad bowl—and mosaic—are in. *Pluribus* is taking over from *unum*, which is seen as too exclusive. The *unum*, some believe, consisted largely of white males—even (in extreme cases) DWEMs (Dead White European Males). Now both the medium and the metaphor are the message. Is there a second less trendy popular culture metaphor?

Popular culture has many facets, like a diamond, and can be subversive and explosive. Scorn may be mixed with the fun, venom with laughter; it can be wildly comical and deadly serious. Popular culture is at the heart of revolutions that slip in on little cat feet. Those most affected by them—the elite and the mighty—seldom see them coming. Popular culture sees and hears, being close to the people. If the medium is the message, then the reaction might be revolution.

These are my thoughts as I look at a famous 1905 photograph that long bedecked my grandmother's parlor. The photograph, like my grandmother, is British. In it King Edward VII and Queen Alexandra pose placidly in a Windsor Palace garden. Edward wears a top hat and frock coat. Alexandra's posh hat is veiled and plumed. In the background an avenue of trees as orderly as the Cold Stream Guards stretches back to infinity. Behind them stretches the past, a sunlit corridor of order and stability.

The same year that photograph was taken, Albert Einstein published his theory of relativity, positing that reality is a complex of masses and motion. Matter became shapeless, to be described only in abstract formulas. Matter and energy were conceived as distinct but interchangeable. Fusion, fission, and the Atomic Age would follow.

With every new age, there is new popular culture—new ways for the people to perceive, absorb, and react to the changes. In the late twentieth century there were so many dramatic inventions, extensions, and transformations that no single "tag" for the age occurred. Computers, satellites, wonder drugs, spaceprobes, social changes involving gender, race, ethnicity, diet, environment, and a host of other things vied for attention.

New Parameters

If "mosaic" is to be our working metaphor, what are our parameters? To attempt an answer shows just how difficult (if not impossible) it is to set parameters for our chosen title. A parameter (from Greek *metron*, to measure) is a set boundary, an arbitrary constant factor. But just what remains constant in today's world? I published a book titled *Parameters of Popular Culture* in 1978. In the intervening years those parameters have grown and changed beyond believing. Our boundaries have been expanded, upended, electrified, computerized, synchronized, put on-line and in virtual reality. We have seen the enforced joining of the unjoinable.

The stratosphere has been joined by the datasphere; the Interstate Highway by the Information Highway. Consumer technology (computers, video cameras, fax, satellite, television, photocopy) has transformed our social environment and hence our society. Media-fed and media-connected, we are suddenly in a New World that few could have imagined a generation ago.

There is a new name for this infiltration: media virus—the instant introduction of icons, ideas, and images into the culture via the media. Spectacular examples are the 1995 O. J. Simpson murder trial, which put not only the man but the whole justice system on trial; and the death of Princess Diana in 1997, which threatened to unseat the ancient British monarchy. Other media events have crowded our collective consciousness: the Rodney King beating, the Willie Horton ad campaign, the Tonya Harding skating scandal, the Newt Gingrich book advance, Whitewater, Clinton's alleged activities with Paula Jones and Monica Lewinsky—all rated as "top stories of the year." Such events, which are by some standards trivial or transient, can quickly snowball into national, even international, obsessions. Where does news end, and trash journalism and pseudo-events take over?

Douglas Rushkoff writes about such matters in his book *Media Virus: Hidden Agendas in Popular Culture* (Ballantine Books, 1994).

He sees all this as a positive development, allowing people to express their opinions, connect with like-minded strangers, and ask the president questions through electronic mail. He believes it extends the boundaries of popular culture, and ushers in a new way to challenge the elitist capitalistic culture. Children's television and MTV, Rushkoff thinks, are the easiest places to launch countercultural missiles. Is that what we want? Must everything be couched in military terms? Instead of extending our humanity, the new parameters might be confining it.

We must always remember that our parameters are far from universal. Half the people in the world have never made a phone call, and many millions live far outside the overconfident system known as the World Wide Web. With 5 percent of the world's population controlling almost half its wealth and consumption, we have a very distorted view of how the world lives. How will the world's parameters change as the Electronic Revolution advances? Will we see more convergence or divergence? It is one of the many questions we must confront, remembering the world alters as we walk on it.

ENDNOTE

See my *American Mosaic: Refiguring American Culture History* (New York, American Heritage, 1998). It has a series of bibliographical references.

The best recent book about the present directions in this field is *Cultural Studies in Question* (London, Sage, 1997), edited by Marjorie Ferguson and Peter Golding.

PART II: TIME AND SPACE

Time

We all live and work in time, but no one knows just what it is. References to and about time are legion. Few words are used so often and so imprecisely. Some picture time as noble and precious—an image of eternity, our only comforter, the great physician (it heals all wounds and wounds all heels), the soul of the world. The "corridors of time" are mythic, always changing, like a fleeting river, a winged chariot, a dream. The moving finger writes, and we can't change a single word. Perhaps it isn't moving but flying—a bird on the wing.

Many others see time as a negative, even a cruel factor. Time is out of joint, a thief, a maniac, a tedious song, a clock setter, a devouring reaper, an avenger. Time is relentless, a shadow that writes no excuses; ambles; creeps with leaden foot; rages like a savage bull. I shall use short vignettes of long stretches of time, trying to find essential, dominating forces, pivotal persons linked to the popular imagination. I have a song to sing.

I sing a song of cycles and the mystery of time. We do not often see cycles, but they are there. We live in time, but never know when our time is up.

Assuming various shapes, rhythms, and timetables, dimly perceived and seldom understood, cycles may occur in one atomic particle or all the cosmos—a moment or eternity. Day follows night, spring follows winter, growth follows decay, epoch follows epoch. In our end is our beginning.

These are not new observations or ideas. All people everywhere have sensed the role of time, and even tried to humanize "Father Time." The rise and fall of civilizations is a perennial time problem. It is always later than we think.

> But at my back I always hear
> Time's wingéd chariot hurrying near.

And yonder all before us lie
Deserts of vast eternity.

Andrew Marvell, "To His Coy Mistress"

That "vast eternity" stretches back to what anthropologists like to call human history—perhaps seven million years by present estimates. Fossils indicate that we might have achieved an upright posture around four million years ago. Stone tools emerged around 2.5 million years ago. At that time *Homo erectus* was more than an ape, but less than a modern human.

Human history took off around 50,000 years ago, in what anthropologists call our Great Leap Forward. From about 40,000 years ago, we have skeletons of people called Cro-Magnons. Biologically and behaviorally they were "modern humans." How did they live? Did they have a popular culture?

Answers to such questions depend on how we define popular culture. Cro-Magnons produced tools and even jewelry, such as ostrich-shell beads. Did this Great Leap take place because of brain change, climactic shifts, the perfection of the voice box? In Europe the Cro-Magnons used their superior skills and weapons to displace the Neanderthals.

For all these centuries, time was measured by seasons, not clocks. One of the greatest changes in human history came when time could be measured, broken down not only into years, months, and days but into split seconds. Nothing marks our way of life more than what I shall call today's hurry-up time.

Hurry-Up Time

There was a time when we seemed to have lots of time. But that time has vanished. We are overwhelmed with tasks at home, at work, at school—even at play. Do what we will, we are always running out of time. Ralph Hodgson's poetic plea goes unheeded:

> Time, you old gypsy man,
> Will you not stay,
> Put up your caravan
> Just for one day?
>
> "Time, You Old Gypsy Man"

Poets and sages have always noted that the moving finger writes and having writ, moves on. T. S. Eliot warns: "Hurry up please—it's time." Time's wingéd chariot, Andrew Marvell notes, is always "hurrying near." But in the late twentieth century, that chariot isn't hurrying near: it's roaring past. HELP!

Time has become a thief. It has stolen our leisure and may even be eroding our civilization. What was once called the leisure class has become the leisure mass. The time issue crosses class, race, and gender lines, affecting family, friends, and communities—nobody has any time out there. Our search for leisure and play has become a scavenger hunt.

Play has become work; leisure remains a mirage. Today's pleasure seeker is like Dante in the meadows of Paradise—a guilt-ridden stranger in a heaven of his own design. "What are you doing here?" Beatrice asks him. "Shouldn't you be back on the job?"

Harvard professors Juliet Schor and Jeremy Rifkin (authors of *Time Wars,* 1993) have uncovered these startling facts:

- One-third of the people who didn't vote in the 1994 elections "didn't have time."

- One out of two Americans doesn't get enough sleep.
- The average work week for Americans has increased 3.25 hours per week—that's an added full month a year.
- Twenty-three percent of the population gets zero days of vacation.
- Parents spend 40 percent less time with children than was the case a generation ago.
- Postal workers must process one letter every second to meet work standards.
- Many labor strikes (such as a recent one at General Motors) are not about higher wages but excessive overtime.
- The epidemic of stress (doctors use the term "time urgency") costs the American economy $100 billion a year.
- American productivity has doubled since World War II—but the promise of leisure seems farther away than ever.
- Half of working mothers say they don't have time to prepare regular meals for their families.

There is a cruel irony here. For years a whole series of "labor-saving devices," automatic machines, robots, frozen foods, and "instant" packaging promised us unheard-of leisure and free time. What happened? Our homes and shops are cluttered with far more stuff; but as Ruth Schwartz Cowan demonstrates in *More Work for Mother* (1985), today's woman devotes as much time to housework as did her colonial sister over two centuries ago.

Homogenized housework has not given the leisure it promises. The end result, Dr. Cowan writes, is that housewives, even the most comfortable, are doing their housework themselves. Housework has become manual labor. The wife of the lawyer is just as apt to be down on her knees cleaning the kitchen floor as the wife of the garbage man.

Children, meanwhile, are overwhelmed by homework, make-work, extracurricular meetings, test scores, SATs, and class ratings. In their classes, sports, and college admissions scramble, winner takes all. Children's games and outings have become a rat race. Children as young as age three spend as long as eight hours a day at summer camp learning to use computers. Some vacation!

Even Little League baseball can become a blood sport. If the umpire seems to miss a call, the parents may decide to rush out onto the field and beat him up. This happens more and more.

Attempting to be the world's youngest pilot, seven-year-old Jessica Dubroff, sporting a pink "Women Fly" cap, rushed into flying—and dying. She was certainly one of the youngest children to make the cover of *Time* magazine (on April 22, 1996). This was the caption: "Who Killed Jessica? Her shocking death raises questions about how far we push our kids."

But surely the parents strolling along their Information Highway have ample leisure? Not so. Downsizing means loading work on those who manage to keep their jobs. That load is increasing at all levels across the nation. True, machines are generating more paper, more memos, more instructions. Just to work through the day's e-mail, faxes, and cellular phone and conference calls turns us into sorcerer's apprentices. Flooded with information, we can't stop the deluge.

But all this might change, you say, when the time-saving Internet covers the whole wide world. Endless information! Instant communication! Everything right here right now!

Don't count on it. The April 29, 1996 issue of *Fortune* magazine casts doubt on this electronic utopia. The headline of a key article reads:

> It's the End of the Net as We Know It.
> An Industry Analyst Explains What the
> Internet Bottleneck Means to Business.

The vision of the Internet as shared network, author William Gurley concludes, is nothing more than an academia-based egalitarian dream that is inherently unsustainable.

So, what is leisure? Why has it backfired?

Leisure: time free from tension and work; being at ease. Leisure is not just a place away, but a state of mind. To be enjoyable it presupposes either great fatigue or some absorbing interest outside the workplace. Most of our work is no longer physically demanding but is mentally draining; we end up with strained nerves. We need to relax—not rest. So what do many of us do away from work? Put our minds in neutral and let talk radio, TV sitcoms, and tabloid journalism take over. We don't cogitate but vegetate, enjoying the cotton-candy nourishment of entertainment. We recreate Plato's Cave in our own living rooms. This, and not religion (as Marx predicted) has become the opiate of the masses.

Why not get away from the grind—enjoy a vacation? All work and no play make Jack and Jill dull. Genuine pure play is a prime basis of civilization. But even play has become work. Pan Am Airlines used to advertise that "Getting there is half the fun." Pan Am has disappeared. So has the fun.

Norman Lobenz traces the vacation pilgrim in his book, *Is Anybody Happy? A Study of the American in Search of Pleasure* (1986). Hub airports remind us of bus stations in World War II. Going by plane is not "traveling" at all. It is merely being "sent" to a place—not very different from becoming a piece of luggage.

Are *these* the "friendly skies"? Landscape? What can you see from 30,000 feet? Hungry? Grab a few peanuts and a drink in a plastic cup to keep you alive until arrival. Then rush on to the motel, symbol of hurry-up-and-relax America. No place is less any place than a motel.

Or rent a car and take your chances on the truck-dominated interstate—the climax in homogenizing the motorists' landscape. Try to hold out until you reach the next McDonald's. No wonder the hapless child, crammed into the minivan's third seat, asks plaintively, "Are we there yet?"

Don't blame "the system," the boss, or the job. We *are* the system—we accept and abide by its dictates. If there is an enemy, it's us. We repeat and endorse platitudes and truisms. If and when we really want to, we can change them. Many are already doing so, in this time of massive transition and rethinking. Instead of "bigger is best" we might say "small is beautiful." Instead of considering the trends of the stock market we might consider the lilies of the field. We could change "Why settle for less?" to "Less is more." Instead of holding up a single finger after winning the game—"I'm Number One!"—we might imitate Churchill and hold up two, to make a "V" for victory—over our own all-absorbing self-interest. Why depend on Ma Bell to "Reach out and touch someone?" We can reach out ourselves. We can teach our children that winning isn't everything. The important thing is not who wins but how you play the game.

Then, perhaps, we can have new communities and real vacations. They might not take us across oceans but into our own backyard. We could end up like our flowers—blooming where we are planted. We could end up smelling the daisies and unplugging the electric clock.

In the beginning there were no clocks to unplug. If we are to understand the way we live, and the popular culture we have created, we must go back to the beginning. That is just where we shall go now.

In the Beginning

In the beginning, God created the heavens and the earth. The Judeo-Christian version comes from the Old Testament; the Islamic from the Koran; the Indian from the Rig-Vedas; and a thousand others from various other cultures, many of which still have only the oral tradition. The creation myth is one of the most popular and prevalent in existence. It changes with the culture and its popular consensus. Many in today's modern world, wedded to scientific explanations, believe in the Big Bang theory. But who set off the bang? Aristotle's answer: the Prime Mover. We are no closer to understanding creation than he was. But our myth is more complex and impersonal.

Behold, I give you a mystery. Consider the question of time. How long did it take to create our planet, our galaxy, our universe? We now know ours is a very small globe in what seems to be limitless space, going back to immeasurable time. Does evolution explain the incredible changes that have taken place, and are still occurring? There must be some sort of cosmic scheme, pattern, calendar. Our greatest twentieth-century physicist, Albert Einstein, strongly believed in a unified field theory—but could not prove it mathematically. The mystery remains.

Where does humanity fit on the cosmic calendar? There are numerous conflicting answers. One that has gained popular support deals not in millions but billions of years. There have been fifteen billion years of cosmic evolution. We are star-stuff, somehow transferred from matter into life and consciousness. Now, five billion years since we evolved from the stars, we are ready to return to them. We have invaded, and will conquer, outer space.

If we transfer this scheme of cosmic evolution to a single calendar year, to allow us to understand it better, the earth began on September 14. Plant life didn't begin until December 23; and Homo sapiens entered the calendar during the final ten seconds of December 31. We are passing strangers on a tiny fragile planet, lost in a cosmic ocean, vast beyond our most courageous imaginings.

Now we switch from the astronomers to the archaeologists, who have had a field day (literally) in recent years. They have found artifacts and evidence that date our prehistoric appearance. In 1964, a team unearthed the nearly complete skull of a creature called *Homo habilis*, which gave us a new idea of early human appearance. This protohuman flourished some two million years ago. Writing of their work in *People of the Lake* (1979), Richard Leakey and Roger Lewin also describe *Australopithecus*, who survived until about a million years ago, then chose an evolutionary dead-end street and disappeared. But somehow *habilis* adapted to the environment, evolved, and passed genes along to *erectus*. This sturdy creature evolved into modern man and woman—Shakespeare's "paragon of animals," and the central figures of the story we shall tell.

A few years later, another expedition found remains of a female in Ethiopia (they assigned the name Lucy) who existed much earlier, as far back as three million years ago. That wasn't the most expansive guess. Carl Sagan spoke not of millions but billions of years of cosmic change, stretching back to "the edge of forever." His book titled *Cosmos* (1980) became a best-seller, and Sagan produced a television series in which he spoke from a magic capsule adrift in the stage-set heavens. After that he marketed maps, trinkets, and his popular views on a grand scale. Meanwhile, many books on astronomy, archaeology, and ethnography were widely reviewed and discussed. Early humans got a new hold on the popular imagination.

So did the boy pharaoh of Egypt, Tutankhamen, whose tomb had been uncovered by Howard Carter in 1922. Suddenly we were engulfed in Tutmania, which we shall discuss in a later chapter; and a King Tut exhibit broke all museum records for attendance. Steve Martin topped the pop music charts with a song regretting he hadn't put all his money into buying a museum. All this baffled students of popular culture. Just what factors linked King Tut and Steve Martin?

Meanwhile, questions about our more distant ancestors came to the fore. If *habilis* is responsible for *erectus* and then *sapiens*, the carrier and creator of popular culture, how did *habilis* live? Is that the ultimate mystery facing us?

The fossil record provides at least a skeleton key, and the lifestyles of primates and modern-day primitives (the Eskimos and !Kung, for example) give more intriguing clues. We now believe that existence for these earlier species was not (in Thomas Hobbes' famous phrase) "nasty, brutish, and short." They seem to have led an uncompetitive life, with far more leisure than most of us have today. Hunter-gatherers spent little time on the hunt and the gathering. Just what did they do most of the time? If only we knew!

How were these distant ancestors different from the primates around them? Sharing, not hunting or gathering as such, is what made us human, Richard Leakey believes. We are human because our ancestors shared their food and their skills in an honored network of obligation. They wanted to cooperate, not annihilate.

Nature itself obscures the lives of earliest people. Nothing is known for sure about the spread of humanity from the cradleland, or if there actually was a single geographical center. Current thinking points to Africa. Hand axes and stone tools show great uniformity through the Old World, and for long periods of time. About nine-tenths of Homo sapiens' existence was spent in hunting and gathering. We know almost nothing of their "popular culture."

Ecology played a major evolutionary role. Glacial ice began to melt from the northern hemisphere around 30,000 years ago; this moist and warmer climate provided food not only for humans, but for mammoths, reindeer, and bison. Populations increased. The melting of the ice allowed hunters to move up the Pacific coast of Asia and to cross the Bering Strait into what is now Alaska.

Grain agriculture and domesticated animals flourished in the Middle East. Many experts now believe that food production began independently in other places, such as America and West Africa. "Civilized history" was well launched. New evidence continues to pour in, as the sciences of archaeology and anthropology march forward.

History was still blind when our early ancestors stood up on two legs, and began to get not only food but information. They must have sensed, long before they recorded it, that knowledge was the road to survival. There is never enough space inside our heads, caves, tombs, or libraries to capture all knowledge—to stop it from growing, fermenting, confusing, advancing, exploding. How to fit it into a workable, governable society?

An Egyptian genius named Menes reached his answer around 3300 B.C.—it was the theory of absolutism, in which power would be carefully related to divinity, and information could be controlled. This led in turn to the invention of hieroglyphics (sacred picture-symbols) in which information could be recorded and preserved. Stone and clay favored hierarchy, and gave us the remarkable Egyptian civilization.

Thousands of miles away, another people, the Chinese, invented paper (around 100 A.D.) and five centuries later, block printing. Now humans had not only the medium of time but of space. In the Indus Valley other advances were being made, and the term "civilization" was born.

The knowledge explosion continued at an ever-accelerating rate. The Romans put out their *Acta Diurna*, which is considered the first newspaper, but the age of print really began when Johannes Gutenberg adapted movable metal type to central Europe's wine presses, and made universal printing a reality. This would foster such things as the Protestant Reformation, capitalism, individualism—and much of what came to be called popular culture.

For centuries, type was set by hand at approximately the same rate: about one line a minute. Only in 1878, with the invention of linotype, did the rate jump to five lines a minute. The typesetter, perfected in 1960, could do fourteen lines, and the 1964 phototypesetter, eighty lines a minute. In 1966 a computer component raised the pace to 1,900 lines per minute. Only two years later the Mergenthaler Linotron photosetter could complete 15,000 lines in a minute. The speed of typesetting had increased by four orders in only ten years. This is only one example of what has happened in our information explosion—which directly affected our popular culture. How did we get from "there" to "here"?

The greatest achievement—and reason for survival—for Homo sapiens has been plasticity of mind. Over millions of years, we have learned how to change and adapt. Nothing reflects this more clearly than our popular culture.

Before the Neolithic revolution, all culture was popular, in the sense of being for the whole populace; of being informed by mutual interests of working together to survive. The more we learn the more we see that the stereotypic "primitives" viewed as savages (in Tarzan stories, pulp magazines, and countless Hollywood B movies) who attack white Europeans, boil captured missionaries, and speak in unintelligible grunts have little to do with reality. Nor are gorillas savage killers, King Kongs, attacking humans and whole cities. Instead, they are peaceful vegetarians who like to lie in the sun.

In 1967, a native called Dafal of the Manubo Blit tribe in the jungles of Mindanao, a tropical part of the Philippines, stumbled on and frightened a band that had been cut off from civilization for countless generations. The Tasaday were, in effect, Stone Age people, living as hunter-gatherers, who had fled into the forest and retrogressed. Slim, muscular, and monogamous, they lived off wild yams, wild fruit, starch from palms, and fat grub worms dug from rotted logs. Eventually in 1971 they agreed to come out and talk with government officials and scientists. The official report indicated these people were very gregarious and had plenty of leisure time. They had no word for "war" and knew nothing to want, except women for bachelors.

"Our ancestors told us never to leave this place," they told the intruders. "They told us the gods would come if we remained here." Was "gods" the right word?

After spending time with these primitive people, Kenneth MacLeith wrote in the August 1972 issue of *National Geographic* that if our ancient ancestors were like the Tasaday, we came of far better stock than he had thought.

* * *

The basic definition of "primitive" is derived from the Latin *primus*, first. Close cognates are prime, primary, prime mover, and prime minister. Landmark books such as Robert Goldwater's *Primitivism in Modern Art* (1967), Claude Lévi-Strauss's *The Savage Mind* (1968), and Stanley Diamond's *In Search of the Primitive: A Critique of Civilization* (1974) have put a whole new face on the word. Let us reclaim our genuine patrimony and identity, they argue. Let us restore what modernism has lost—making restitution for what the prodigal son has done.

Although the 1889 Paris World's Fair and the work of van Gogh and Gauguin featured primitivism, it was Picasso's turning to primitive African art to invent cubism in 1907 that has given primitivism a central role in the modern world. *Les Demoiselles d'Avignon*, completed that year, is generally acknowledged to be the first great cubist painting. It has lost none of its power and impact in nine decades.

What was involved was a radical idea that changed the history of Western art—that the tradition of the human figure, which was central for over 2,000 years, has at last run out. To renew its vitality, we must study and adapt the primitive distorted forms of African art. "Its lashing rhythms," Robert Hughes writes in *The Shock of the New*, "remind us that Picasso looked to African masks as emblems of savagery, violence transferred into the sphere of culture."[1]

Since then the "primitive impulse" has played into our romantic, emotional, intellectual, and subconscious thinking. This in turn goes far in helping us interpret late-twentieth-century popular culture and what is called postmodernism.

My comments on the long and often dim history of popular culture make no attempt to be all-inclusive, scientific, chronological, or strictly thematic. Mine will be a personal interpretation, growing from years of teaching, traveling, and writing, and will move back and forth in space and time. Most of my material will be about recent centuries and Western civilization. It will adhere to one central theme: the parameters of popular culture need to be expanded and explored. I hope to refute those who think popular culture is a newcomer to the human drama; that it is only electronic; or that it sprang up in the 1950s with the beatniks, hippies, new journalists, pop artists, and celebrities such as Andy Warhol, Elvis, and Marilyn Monroe.

Popular culture is as old as humanity itself. We know all too little about it and its impact on history. Culture has always been "popular," thriving on formula, archetypes, and stereotypes, as well as fads and follies. It is circular, repetitive, and powerful. Here today and gone tomorrow—but just wait, and it will be back next year, next decade, or even next century. And—looking backward—it may have been "popular" long ago. We like to think of ourselves as inventing mass production. Yet the Sumerians were mass producing funeral effigies centuries ago; so were the Egyptians, Tibetans, and Chinese.

We may be in a new Space Age—but we also seem to be neo-Victorians in many aspects of our lives. The French had it right: *Plus ça change, plus c'est la même chose*. The more things change, the more they remain the same. Yet we never know just when, why, or how change will come about.

ENDNOTE

Few fields have seen such dramatic breakthroughs and advances as archaeology. The books below all contain excellent bibliographies of earlier standard works. These four are of special note:

1. Michael O'Brien. *Evolutionary Archaeology: Theory and Application.* (Salt Lake City, University of Utah Press, 1996).
2. Colin Renfrew and Ezra Zubrow. *The Ancient Mind: Elements of Cognitive Archaeology* (New York, Cambridge University Press, 1994).
3. Brian L. Molyneaux. *The Cultural Life of Images: Visual Representation in Archaeology.* (London, Routledge, 1997).
4. Sian Jones. *Archaeology of Ethnicity: Constructing Identities in the Past and Present.* (London, Routledge, 1997).

Cavespace

On a July afternoon in the summer of 1879, amateur archaeologist Marcellino de Sautuola was digging in the rubble near the mouth of a recently discovered cave called Altamira, near Santillana del Mar on the north coast of Spain. His nine-year-old daughter, Maria, who had gone into a side chamber, yelled "*Toros! Toros!*"

Marcellino dropped his pick and rushed in. When he raised his lantern, he later said, he felt his hair stand on end. The ceiling was crowded with seventeen gigantic paintings—not *toros* (bulls) but prehistoric bison, magnificently painted in shades of red, brown, yellow, and black. They were standing, bellowing, rolling in the dust, dying of wounds. There were other animals—a horse, a trembling cow, a wolf, and wild boars.

That was only the beginning. Moving farther into the cave, he discovered paintings of giant aurochs, great-antlered stags, a cave lion, and a woolly mammoth. These were animals that were either extinct or had disappeared from Western Europe centuries before. He was looking at the primitive equivalent of the Sistine Chapel. On that one incredible day, he had opened up a totally unsuspected era of human history, and identified a new benchmark in what would later be called popular culture.

As is often the case when great breakthroughs occur, he was discredited by the Establishment. A congress on prehistoric archaeology, meeting in Lisbon in 1880, rebuffed his findings. How could such sensitive art be done by savages not far removed from the ape? Had Sautuola painted the pictures himself? Was it a "Jesuit plot" to discredit the new science and evolution? Crushed by this rebuke, Sautuola died in 1888, his honor impugned, an object of ridicule.

History has exonerated him. In recent years we have come to understand that Homo sapiens began its art career by representing large animals in caves and shelters. More than 200 such caves containing thousands of images have been found in Italy, Switzerland, France, and Spain. The oldest, believed to be over 30,000 years old, is the cave at Chauvet, in France's Ardeche River valley. The most recent cave galleries are now thought to be about 10,000 years old.

No cave yet discovered enjoys the reputation and close scrutiny of the best known and publicized cave—the one that appears most often in art books and histories—near Lascaux in southern France. In 1940 schoolboys slid down a rocky chute, to end up in what is now sometimes called "the Louvre of Paleolithic art."

The Lascaux cave is indeed magnificent. Like many other pilgrims, I have stood speechless before its vast array of animals, some up to eighteen feet long, with sixty horses, twenty oxen, a wounded bison with a lance across his hindquarters, and a whole array of prehistoric beasts. The artists used natural earth colors, yellow ochre, and lampblack made from animal fat. A hollow bone possibly used as a tube has been found in the cave. The workmanship and imagination are superb.

Standing in Lascaux, looking at this splendid triumph of the human spirit, we ask: Has art really improved much since then? Have we made any more profound portrayal of our constant struggle to survive in a precarious world?

The Dordogne region of southern France is particularly rich in cave art. Font-de-Gaume, for example, has over 200 pictures covering many periods; Les Trois Freres has over 500.

Why was this art, the foundation of popular culture, created? While we can never answer for certain, we have clues. The most persuasive explanation came in the early 1900s from Abbe Bruel, who thought the paintings were involved with hunting magic. By "killing" the animals' images, the hunters thought the same might happen when the outdoor chase began.

Later scholars have accepted and elaborated on Bruel's "hunting magic" theory. Perhaps caves had magical significance as symbols of the womb. Three kinds of magic have been suggested: (1) fertility magic, to create new animals and food; (2) death magic, to allow hunters to kill their prey; and (3) propitiation magic, to appease the dead animals, and ensure the continuance of the hunt.

Painted caves were also believed to harbor demonic spirits. At some early unknown date Christian monks painted crosses over cave paintings in the Lake Onega region in what later became the Soviet Union. Here too we encounter a form of magic.

Art is magic. People everywhere have believed that. As Edward O. Wilson points out in *Consilience: The Unity of Knowledge* (1998), hunting magic and sorcery still exist in hunter-gatherer societies in our own day. Prescientific people believe that by manipulating symbols and images one can actually influence reality. In today's popular culture, voodoo doctors stick pins in dolls. Many Americans still carry a rabbit's foot in their pocket.

Children chosen for sacrifice to the Aztec god Tlloc were forced to shed tears, which would bring rain to Mexico. For Christians, baptism washes away sins. To be "born again," you must be "washed in the blood of the Lamb." Students of popular culture know that people seek power and forgiveness by any means they can muster. How many millions still play the lottery?

What about humans in cave paintings? There aren't many. Most are shamans wearing animal skins or horns. In other places, such as the Fertile Crescent and Mesoamerica, animals appear as gods. Thousands of sacred mummified cats have been found in Egypt. Today we worship our animals in the names of sports teams—lions, bulls, bears, tigers, dolphins, and eagles.

We marvel at the skill of the early cave painters. The paintings are meticulous and convincing with clean sweeping lines. Many are accurate and beautiful by any standard. The paintings display a whole range of mammals, from lion to mammoth, rhinoceros to bison, male and female. Some females are pregnant. Arrows fly and animals die. At Chauvet two fierce male rhinoceroses fight to the death.

Recently, decorated caves have been found around the world—in Asia, Africa, and Australia. The more we learn the more we are awed by this early form of people's art—popular culture.

Such art flourished with isolated Eskimo tribes in the frozen North, the Tierra del Fuego Indians at the southern tip of South America, the pygmies in African forests, and various aboriginal tribes in Australia. Recent work on the last of these groups has uncovered remarkable poetry, such as this excerpt from the Dulngulg cycle of Australia's Mudbara tribe:

> The day breaks—the first ray of the
> Rising sun stretching her arms.
> Daylight breaking, as the sun rises to her feet.

Might the cave people who produced their splendid paintings many thousands of years ago have had their poetry, folktales, drama, riddles, jokes? Did they produce epics many centuries before Homer produced his in Greece?

N. W. Thomas, whose book *Natives of Australia* raised these questions as early as 1906, produced no answers. Perhaps we will never have any. All this reminds us of how many chapters in the history of popular culture may remain forever blank.

The word "primitive" has negative overtones—instead of its original meaning, "first." We should confront every culture within its own cultural system, not by comparing it to ours. Until we do that, we shall misunderstand the world. Primitive art, be it in early Lascaux or today's remote African forests, expresses the distinct and sacred feeling of people who occupy the same world, confront the same life-and-death decisions, that we must make.

In *The Spirit of Folk Art* (1995), Henry Glassie points out how outmoded ideas of race or class can cause us to reach disastrous conclusions:

> It cannot be that we think fine art is the art of pale wealthy people, folk art is the art of pale poor people, and primitive art is the art of dark people. The world is too grand to be splintered into bits for scholastic convenience.[1]

A primary job, if we would understand either folk or popular culture, is to learn to understand and value what our own culture devalues. That is a formidable task, but in the days ahead, we must undertake it.

For many centuries folk art and popular art were synonymous. Only in the Western world—and especially in the United States—have we tended in this century to separate them, linking "popular" with the electronic age and mass production. What this "mass" has done, notes Henry Glassie, is to strip the workers of individual excellence—their products have no relation to art. He concludes in *The Spirit of Folk Art:*

> Instead of a chisel or a brush, the designer uses people who use machines to drag out of a massive public a small group who understand his effort.[2]

Art is made to be used—the icon to venerate, the amulet to thwart witches, the pot to carry water. When the utility is gone, so is much of the power. When an African carving is no longer useful, it is discarded—or given to tourists. When a piece of popular art is useless—like a plaster of paris hamburger—it might end up in a museum. Don't try to eat it.

* * *

These superb visual art achievements, not only in Europe but throughout the world, were the work of people we call hunter-gatherers. In remote and isolated places, hunter-gatherers still live as the ancient cave dwellers might well have done.

Do hunter-gatherers everywhere—then and now—share any common tools, fears, beliefs? We believe they do. First and foremost, all have creator gods, supplying myths of creation. All have weapons, burial tools, ochre to color the dead, fertility rites, mother-goddess carvings, and images of animals associated with magic. All shared and share a deep, pervasive respect for the spirit world and ghosts. Long removed from the caves, and from our food-gathering days, modern societies still contend with many of these vestigial remains. Sigmund Freud and others revealed the many psychoses and neuroses that beset our times. Fear stalks our streets and our homes. Now we don't go to shamans or witch doctors to seek relief. We take Prozac.

The more we study the past the more we realize it isn't really past. Instead, it is buried just under the surface, waiting to emerge and explode in the twinkling of an eye.

* * *

We know more about the next great change that occurred in human history, labeled the Agricultural Revolution. Forms of work changed, stable locations developed, people finally had some control over nature. New social classes, new concentrations of wealth, and urban centers sprang up. With city life came extended trade, government, organized religion, schools, writing, and records. These developments, taking place from about 4500 B.C. to 3000 B.C., were centered in the Nile Valley, the Tigris-Euphrates Valley, and the Indus Valley. We find our first evidence of the wheel from about 3500 B.C., as well as the smelting of copper—the first metal innovation. This in turn gave people new power over other materials, such as wood and leather. Yet much remained from the earlier hunter-gatherer life—birth-to-death rituals, floods, famines, disease. All of them are still with us today.

* * *

But the most intriguing cave question takes us to Greece, and a great philosopher who is surely a founding father of not only Western civilization, but also popular culture—Plato. His allegory of the cave, in his treatise *Republic*, is as relevant and challenging today as it was 2,500 years ago.

So extensive is the influence of Plato (427-347 B.C.) that one tends to agree with Alfred North Whitehead: all of Western thought is a series of footnotes on Plato. A wealthy aristocrat, he witnessed the execution of his beloved teacher Socrates in 399 B.C. After travel abroad, Plato founded his academy in 386, when he was forty-one. It became the model for what we now call the "academic way of life." Ten years later Plato wrote *Republic*—a statement of the academy's aims and goals—and the famous allegory (some would say parable) of the cave.

In it he imagines a group of citizens trapped inside a cave, unable to see anything except the wall in front of them. Behind them is a fire, and figures are passed before it so as to cast shadows on the wall. What they see, of course, is the mere image of reality; but, unable to look back, they think they are seeing reality itself. They are, in short, living in a world of illusion.

The startling thing is that they enjoy it. When someone comes from the outside and offers to release them, and take them out of the cave, they resist. If they are taken outside, the sunlight blinds them, and they want to reenter the cave. They prefer unreality to reality. The job of the teacher, says Plato, is to convince them that reality is better—then to go back and release other cave prisoners.

Applied to the caves we have been describing, the message is obvious: one can't live away from the world, confronting animals only in wall paintings, and survive. More compelling is the thought that today's cave is our movie house, TV room, or computer desk. Here too we gaze at images of reality, not the real things themselves. Turn off the power, and the film fades, the TV screen goes blank, the computer loses its image. But—like the reluctant cave dwellers in Plato's dialogue—we can be seduced by the false images to prefer them to reality. We can live in a fantasy land, content to let the rest of the world go by. Popular culture has invented a word for such victims: they are "couch potatoes." We shall visit them again in a later chapter.

Now we have come to realize that it is not only on the couch, or in the darkened movie house, that we are "in the cave." The whole culture can trap us as well. We can be so narrow-minded, so chauvinistic, that we can't even imagine life elsewhere. We laugh at Mark Twain's saying "In France even the children speak French." But how funny does it seem to the French?

Or again: might some phase of our culture—elite, folk, popular—cut us off from others? The cave metaphor operates on several different levels and contexts. In the twenty-first century, will popular culture tend to be one of the major ways for the human race to become more tolerant, more universal, more open-minded? Will the operative word not be local or national, but global? We shall confront that question later on. But for now, we move to one of the most stable and influential cultures in world history—Egypt. For thirty generations that pivotal land maintained a remarkable conformity in its religion, art, and lifestyle, which can still attract and entrance us today. It is time to speak of Tutmania.

ENDNOTE

Books and manuscripts dealing with the thousands of years this chapter covers number themselves in the thousands. The single book that has helped me most—and which I quote twice—is Henry Glassie's *The Spirit of Folk Art: The Girard Collection at the Museum of International Folk Art* (New York, Harry N. Abrams, 1995). The leading folklorist of our time, Glassie shows us a new way to look at the past, and define our humanity.

Another bold and rewarding recent book is Hugh Thomas's *World History: The Story of Mankind from Prehistory to the Present* (New York, HarperCollins, 1996). Earlier classics include Pierre Louis Roederer, *Oeuvres*, 8 vols. (Paris, Pleyel, 1853-1859) and

Lewis Mumford, *The Transformations of Man* (New York, Macmillan, 1956). Two recent books that include the latest information, presented in an excellent style, are Edward O. Wilson, *Consilience: The Unity of Knowledge* (New York, Knopf, 1998) and Jared Diamond, *Guns, Germs, and Steel: The Fate of Human Societies* (New York, W. W. Norton, 1997).

Tombspace

Egypt: mystery, magic, pyramids, tombs, buried treasures, Cleopatra (69-30 B.C.), passionate love that wrecked empires, secret symbols and ceremonies. No human culture has been more popular than Egypt's. Half of recorded history had passed before anyone in Europe could read or write, but the priests of Egypt were keeping accurate records between 4000 and 3000 B.C. And artisans there were creating objects still held to be among the most beautiful ever created.

Strategically located at the meeting of Asia, Africa, and Europe, a prize for any world conqueror, Egypt has had a peculiar and enduring fascination for centuries. One of her boy pharaohs, whose tomb was discovered intact in the early twentieth century, has made a huge impression on late-twentieth-century America. We still suffer from a deep-seated case of Tut Fever.

The essential story of why Egypt still figures prominently in our history, and why tombspace moves us forward beyond the era of cavespace, has to do with a major information revolution. A buzzword of our time, linked to computers and satellites, it goes back five millennia to an Egyptian genius named Menes.

Founder of the First Dynasty of historical kings of Egypt, Menes was long looked upon as a mythical figure; but the discovery of tombs, including that of Menes himself, has placed him firmly in history.

The leader of the falcon clan, which united northern and southern Egypt around 3300 B.C., Menes moved the capital to Thinis and created a kingship to meet the demands of a uniform system of religion and administration. One result was the invention of hieroglyphics—sacred picture writing.

Many think Menes was the father of monotheism; all agree that (as Harold Innis points out in *Empire and Communication,* 1972) Menes established a state religion with elaborate temple endowments, rituals, priesthoods, and a cult of the dead. He developed the theory of the absolute power of kings, which in turn strengthened the idea of immortality. Mummification carried out the idea of pictorial representation as an essential element of funerary ritual.

As time went by, figures became definitions of living beings and objects. The king or pharaoh was a sun god who did not work with his hands, but merely existed. Like the sun, he acted on the environment from a distance. An order of the king was seen as equivalent to an act of creation. His command was reinforced by the mystery of writing as a reflex of the spoken word. Thus did philosophy, religion, and politics overlap and merge. All this helped to create popular culture, flourishing for thousands of years and continuing up to our own day.

This peculiar and enduring fascination was manifested with the early Greeks and Romans, who felt that their civilizations were new and inexperienced compared with Egypt's, where the past was strangely alive in the present. The Greek Herodotus, known as the "Father of History," felt this way, as did many who followed him.

And as we do, profiting from all our modern archaeology, anthropology, and history. The fourth millennium, in which civilization took shape in Mesopotamia and Egypt, was the most fundamentally creative era in human history. When that millennium opened, the Near East was a mosaic of Neolithic farming communities, nomads, and people entirely dependent on hunting and fishing. By 3000 B.C. the entire Nile Valley was organized into a single state, and the artistic and architectural achievement was impressive. The basic social structures—and popular culture—that would characterize all later civilizations up to the Industrial Revolution were in place. Most of the people were farmers or fishers, controlled by a tiny elite of landed nobles and priests. At the top of the power structure was a king, pharaoh, or emperor (the name varied, but the authority was always supreme) who claimed some special kinship with the gods. This pattern was followed in the East and the West, the Old World and the New.

Many have condemned this pattern, with its exploitation, and it clearly imposed a tragic burden on much of humanity. Karl Marx based his new socialism on the upending of class exploitation, and Sigmund Freud even blamed it for many of our basic neuroses.

Nevertheless, this ancient hierarchical system, with all its flaws, opened up venues of creative human expression that were unknown before. It provided large-scale political and economic organizations that increased food production, community spirit, and security. They were far less prone to death and disaster than the earlier hunter-gatherer and primitive pastoral societies, which were always in bondage to the whims of nature. Theirs had been, for countless centuries, a world of desperate insecurity and dark terror.

The Egyptian advances in mathematics, engineering, temple construction, road building, and administration were extraordinary. The Near East also shaped the religious concepts that would be central to three great world religions—Judaism, Christianity, and Islam. Protected by deserts surrounding their Nile Valley, the Egyptians were able to create a rich civilization that still fascinates and challenges us. A broad-based study of popular culture must begin here.

At the heart of Egypt's special fascination is the Nile—a slow river with long gentle floods. Each year it brings down fresh soil (silt) which provides abundant crops, enough to sustain the Egyptians and export the surplus to many other areas.

Egypt's wealth sustained lavish engineering projects, including the Great Pyramid of Cheops, completed in 2566 B.C. Using six million tons of stone, which took ten years to quarry, it remains a leading icon and tourist attraction throughout the world.

Over the years, Egypt changed from simple village life to estate and city life—from a classless society to one resting on a strict earthly and heavenly hierarchy, with a god-king called Pharaoh. Two of William Shakespeare's most popular works (*Julius Caesar* and *Antony and Cleopatra*) play on these themes. Egypt has long stood for mystery, gods sent down from heaven as pharaohs, builders of magnificent tombs and pyramids, style-setters that lasted throughout the centuries.

Could popular culture ask for anything more than that?

Invaders were always impressed by the high culture, wealth, and obsession with death of the Egyptians, who conceived of death as a continuation of life in another realm, and buried their leaders with food, clothing, money, and beautiful objects.

The recurring invaders—Persians, Greeks, Romans, Arabs, Turks, Europeans—all found this remarkable, and fell under the Egyptian spell. They carried off what they could, then came back for more. In modern times, no museum worthy of the name could be without Egyptian relics.

Writing in *Egypt As It Is* (1877), J. C. McCoan noted that up to 1872, mummy bones ranked high on the list of Egyptian exports, with these bones contributing nearly as much as those of modern cattle to the yearly total of 10,000 tons, sent chiefly to England.

The discovery of a treatise called *Egyptiaca,* written by a priest named Manetho about 240 B.C., listing thirty dynasties of king and pharaohs, increased the sense of Egypt's antiquity and eminence. This sense was further displayed in several of Shakespeare's plays, Napoléon's colorful conquest of Egypt, guidebooks, and a whole series of popular best-sellers, such as J. G. Wilkinson's *Manners and Customs of the Ancient Egyptians* (1837). Museums around the world paid high prices for Egyptian antiquities. Demand created supply and tomb robbing became a lucrative occupation.

More and more attention centered on tombs and mummies. There were scores of them, stretching back centuries. Little wonder. As soon as a new king or pharaoh came into power, he selected a site for his tomb, his "house of eternity," and began work on it. It usually contained two parts: the underground burial chamber and the outer rooms. Life continued for the king after his earthly death. He lived in the world of the stars, ascending to heaven to join the gods.

His lavish tomb contained gold, furniture, chests of clothes, food, and weapons. Sometimes slaves were slain and buried at his side so they might serve him in the next world. In Middle Kingdom tombs, whole groups of figures worked in model kitchens or workshops. There was in addition an endowed estate and a "ka-priest," or "priest of the double," endowed for perpetuity.

* * *

The fascination with ancient Egypt carried over into modern Europe. It was particularly strong in Germany and England. One of the many English aristocrats involved was Lord Edward Stanhope Molyneux Herbert, fifth Earl of Carnarvon, a sporty gentleman who declared he would "rather find an uninvaded pharaoh's tomb than win the Derby." For eighteen years he underwrote annual expeditions to

Upper Egypt, in the Valley of the Tombs of the Kings, hoping to find a lost, but perhaps intact, tomb of an Egyptian pharaoh.

His chief assistant and field marshal was Howard Carter, a self-taught Egyptologist and lonely bachelor whose entire life was given over to Egyptology. Carter, who would be the discoverer of the most famous tomb in modern times, had had a makeshift career. In class-conscious Britain he was not "upper-class." But he did have perseverance. He wanted to discover and explore the original tomb of a great pharaoh. He learned to speak Arabic and Egyptian dialects, then went to live and work with the natives. He built a permanent home in Egypt's Valley of the Kings, and got to work.

He needed and got a wealthy sponsor, Lord Carnarvon. On and off for eighteen years, Carnarvon financed Carter's work. With this backing, Carter hired a hundred men and moved thousands of tons of chippings—no luck. By 1922, Carnarvon was ready to give up; Carter pleaded for one more try. On November 1 he unearthed a step that seemed to lead to a tomb. Carter wired Carnarvon, who hurried to Egypt with his daughter. At last, a real breakthrough. Then, on a single memorable day, November 26, Carter's years of perseverance paid off.

He later described it as "the day of days, the most wonderful I have ever lived through." He dug a hole into what he hoped was the tomb of a legendary pharaoh named Tutankhamen. He widened the breach and inserted his candle, which continued to flicker. The next time he poked his head in. At first, he saw nothing. An eternity passed. Then he was "dumbstruck with amazement." Carter recalled the moment in his own diary entry:

> But presently, as my eyes grew accustomed to the light, details of the room within emerged slowly from the mist, strange animals, statues, and gold—everywhere the glint of gold.
>
> His anxious sponsor, Lord Carnarvon, asked, "Can you see anything?"
>
> "Yes,"Carter replied, "wonderful things!"

Wonderful, indeed. He had discovered the tomb of King Tutankhamen—the first tomb of an Egyptian pharaoh ever found virtually intact, which would launch one of the most incredible sagas in the history of popular culture.

Archaeologists had known about King Tutankhamen since Napoleon's day. He was a phantom pharaoh. No one could tell what his lineage was, where he was born, how he died in his adolescence. The one point scholars agreed upon was that he ruled Egypt at a crucial time in her development.

His great challenge had been to stave off the growing power of a middle-class bureaucracy and win back the forsaken gods. In this he succeeded, returning to the worship of Amon, restoring to the gods their old privileges, and reinstating the private worship of Osiris. He built new ships and covered them with gold. He won the people's hearts.

Benevolent and popular, he surrounded himself with slaves, singers, and tumbling girls. His people were delighted, and his popularity increased.

The newly discovered tomb, undisturbed for 3,500 years, was a treasure trove. When the announcement was made, bedlam broke out. Writers, reporters, and visitors poured in from around the world. Only three weeks after the announcement of the discovery, the first of a whole string of instant potboilers appeared: Frederick M. Burns' *Tut-ankh-amen, or the Valley of the Kings* (1924). Burns promptly copyrighted the use of the name Tutankhamen on dozens of items, ranging from dolls to bracelets to cigars. Other promoters moved quickly to copyright such names as Tut, Tut-Tut, and Two-Tank. An enterprising American businessman offered $100,000 outright plus royalties for the use of designs found in the tomb. So-called "relics" supposedly from the tomb went on sale in New York. The situation got so bad that the New York State Assembly passed the Tut Relic Bill, making it a crime to sell bogus relics. The bill is still on the books.

Meanwhile, a bevy of chorus girls in "Tutankhamen bathing suits" paraded around Miami Beach, and belly dancers shook from coast to coast. President Harding sported a Tut Cane. Every state fair featured a "Little Egypt," and Vincent Lopez, a popular bandleader, began to play "Tutankhamen music." A German filmmaker started work on *The Pharaoh's Revenge* and Hollywood produced *The Mummy*, featuring tombs, pith helmets, mummies, Egyptologists, and a beautiful blonde heroine. A star was born for the new genre—Boris Karloff—and he rose to fame in the next generation.

By then Tutmania was here to stay. Tutankhamen no longer belonged just to Egypt, but to the whole world. Soon the world would be caught up in a new disease—Tutmania. It has never gone away.

* * *

But the most intriguing part of the whole story we have yet to mention: the Mummy's Curse. For the student of popular culture, that is where "the moral" rests.

Those who believe in the curse date it from the time Carter discovered the tomb; on that day, a cobra appeared and killed his yellow canary. The Egyptian workmen thought the yellow bird was responsible for the discovery, and spoke of "The Tomb of the Yellow Bird." Now the bird was being punished.

Not only the canary, but Lord Carnarvon himself. He was bitten by what he thought was a mosquito, but his health deteriorated so rapidly that his family was called to his side. On April 5, 1923, at age 57, he died, less than twenty weeks after viewing the mummy's resting place.

At that very moment Cairo's electric power failed, and the capital was in total darkness. No one was able to explain what caused the outage. Then a severe plague broke out in Upper Egypt.

All this seemed coincidental, until a distinguished French Egyptologist, Professor J. S. Mardrus, attributed these strange events to "incomprehensible occult powers." Those who violated the tomb would be cursed.

New evidence accumulated. Multimillionaire George Jay Gould visited Luxor, was stricken down, and died. So did Sir Archibald Reid, famous radiologist, when he signed an agreement to X-ray Tutankhamen's body. The well-known Egyptologist H. G. Evelyn White committed suicide "under bizarre circumstances." The most frightening tale of all emerged when Woolf Joel, a close friend of Carnarvon, en route up the Nile to Luxor aboard Carnarvon's yacht, was mysteriously swept overboard and drowned. How did Professor Mardrus respond to all this? "I am, unfortunately, not at all surprised."

But everyone was surprised when Lord Carnarvon's half brother, Aubrey Herbert, died of "temporary insanity"; and Lady Elizabeth Carnarvon, just like Lord Carnarvon, died of an insect bite.

In 1929, Howard Carter's personal secretary, Richard Bethell, died of an undetermined cause in London's Mayfair Club. This so upset his father, Lord Westbury, that Westbury committed suicide.

The hearse in which Westbury's body was placed ran over and killed an eight-year-old boy. By 1935, the number of attributed "victims" of the curse reached twenty-one. "It's a deep mystery," *The New York Times* concluded, "which it is all too easy to dismiss by skepticism."

Almost seventy years later we still confront what many believe is "a deep mystery." Arthur Conan Doyle, creator of Sherlock Holmes, attributed the mystery to "elementals—created by Tutankhamen's priests to guard the tomb." Scientists frame their answer in more scientific terms: infection by long-dormant microorganisms (histoplasmosis).

But how to explain these events: Carnarvon's young brother died suddenly in 1923; an X-ray specialist died unexpectedly en route to examine King Tut; a reporter died of pneumonia after visiting the tomb; the Egyptian Ali Kemel Fahmy Bey was shot by his wife after viewing the tomb; French Egyptologist George Benedite died as a result of a fall after peering into the ancient space. No one can deny that death was peculiarly selective in its choice of victims.

One idea about the genesis of a curse rests in the sudden death of Lord Carnarvon. However, there is a scientific explanation for this. Almost killed in a 1901 automobile accident, his lungs were terribly damaged and weak. After the insect bite, he got blood poisoning, had no modern medication, and died.

Skepticism, mysticism, fanaticism—folklore or fakelore? Here is a question that requires more research and the use of many different sources. One persistent rumor has it that Howard Carter found this inscription in the tomb: "Death shall come on swift wing to him that toucheth the tomb of the pharaoh." No such inscription has ever been found.

The whole matter took on new life in 1976 when the Egyptian government agreed to send "The Treasures of Tutankhamen" to America for the Bicentennial celebration. It was a major cultural event of huge significance. The great obsession over King Tutankhamen reached epidemic proportions. "Tutmania" was the operative nickname. Others called it "Tut glut," "Mummy madness," and "the Egyptian fever." Stores ran out of their "Love My Mummy" T-shirts; everything and everybody glorified King Tut, the boy idol.

To accommodate as many visitors as possible, the Metropolitan Museum in New York stayed open seven days a week, for a total of eighty-two hours, allowing well over 1,300,000 to crowd through its doors in two weeks. People camped out, got mad, finally got in—it was the same story in New Orleans, Chicago, Seattle, and Los Angeles. There had never been such interest in an art exhibit. Those who couldn't see the exhibit could watch Steve Martin on television, doing his new King Tut song and dance.

* * *

Scholars argue that tomb culture covers vast and important findings, and the King Tut find has been blown out of all proportion. From their viewpoint, that may be so—but in the realm of popular culture, no other discovery has had nearly the impact. Sir Leonard Woolley's discovery of Abraham and Ur, Layard's discovery of Nineveh (the city of Jonah), Koldewey's discovery of Babylon (the city of Daniel), and Schliemann uncovering the site of the Trojan War may be more important than anything Howard Carter did. But Carter has become a sort of folk hero, who stumbled into immortality—the poor boy who made good against incredible odds. With him, Horatio Alger rides again.

* * *

Tombspace has proven to be most spectacular in the Middle East. The Sumerian tombs of Ur are of major importance. *Gilgamesh,* the first written masterpiece of world literature, was found in a tomb of a Sumerian king of the city of Erech (Uruk). But major tombs have by no means been confined to that area. The Indus Valley has magnificent tombs, in and around such great cities as Mohenjo-Daro and Harappa, destroyed by Indo-European invaders around 1700 B.C. Magnificent Mayan tombs have been unearthed in Central America, and work has begun on tomb sites in Africa.

No modern tomb has received more veneration than that of Lenin, in Moscow's Red Square. Millions of people have gone there to see the well-preserved body of a founder of the USSR. Even when the nation crumbled, the tomb and the body continued to be popular. Tombspace has been sacred space for thousands of years and will always remain so. It is the way we pay homage to our most fearful foe, death.

ENDNOTE

The enormous library on Egyptology has added many new titles since the King Tut Glut. I have found these four especially useful:

1. Nicholas Reeves. *The Complete Tutankhamen: The King—The Tomb—The Royal Treasure.* (London, Thames & Hudson, 1995).
2. Elaine Landau. *The Curse of Tutankhamen.* (Brookfield, CT: Millbrook Press, 1996).
3. Elaine Edgar. *A Journey Between Souls: The Story of a Soldier and a Pharaoh.* (Lafayette, CO, White-Boucke, 1997).
4. Bob Brier. *The Murder of Tutankhamen: A True Story.* (New York, Berkley Publications, 1998).

For the moviegoers, there are lots of mummy movies to choose from.

Templespace

One can fly across the Mediterranean from Egypt to Greece in an hour—but the cultural differences stretch over millennia. Egypt was obsessed with permanence, authority, continuity; Greece with expansion, experiment, individuality. The Egyptian pyramid—heavy, dark, built for the dead, makes a startling contrast with the Greek temple—open, light, built for the living.

The most famous temple in Greece and in Western civilization sits atop the Acropolis in Athens—the Parthenon (447-432 B.C.). This unique masterpiece has come to epitomize the intellectual and artistic glory of Western civilization. The Parthenon is both a temple and an icon; like only a few other structures in the world, both landmark and mindmark. Think of the Pyramids, the Great Wall of China, Angkor Wat, the Taj Mahal, Chartres Cathedral, Versailles, the Eiffel Tower, the Empire State Building. Peaks of human achievement, they are perennially popular.

The Parthenon was the crowning achievement of Pericles, mastermind of Athenian democracy and its empire. Sympathetic with both aristocratic and democratic elements, he was a believer in the sovereignty of the people and the role of popular culture. The pivot on which the constitution turned, Pericles was elected annually as Athens' leader for fourteen successive years. His strength and power created Athens' templespace.

Two of the most talented citizens, Ictinus and Callicrates, were chosen to be the architects of the Parthenon, working under the general supervision of the sculptor Phidias, who provided the pediments, metopes, and friezes, as well as the gold and ivory statue of Athena inside the temple. These extraordinary artistic refinements combine to give the Parthenon the majestic sculptural appearance that made it both a shrine and a symbol.

Is any building in the world better situated? Built of white marble, perched on the southern flank of the Acropolis against the backdrop of mountains to the north, the Parthenon dominates the entire city and countryside. Covered with precious stones and statues, it was and still is a wonder to behold.

Not one but four different cultures utilized it. In Christian times, the Virgin Mary replaced Athena. The temple first became a Byzantine shrine, then a Catholic cathedral, and finally a Turkish mosque. A ruin since 1687, destroyed by Turkish gunpowder, there was still one more dramatic chapter in its history. Much of the sculpture was removed between 1801 and 1803 by Lord Elgin, and put on display in the British Museum. Now crowds of visitors go there to see this incomparable treasure of our ancient past.

If London alone has these precious remnants, cities all over the world have temples serving not only as religious shrines, but state buildings, banks, factories, and railroad stations. The idea of the temple with columns adding both strength and dignity has now become symbolic of the world's heritage and architecture.

* * *

In his popular and influential book called *Towards an Architecture* (1923) the Swiss architect Charles-Éduoard Jeanneret (Le Corbusier) chose the Parthenon to celebrate what he called the "White World"—one expressing clarity and precision, exact proportion and precise materials, and profound beauty of a building and the people who erected it. It was animated by a single thought, drawing around it the desolate landscape and gathering it into the total composition. That thought, so influential on so many architects and planners, has carried templespace into public buildings, churches, banks, and factories around the world. "Modern" architecture, with its sparse clear lines and lucidity, ties the ancient temple with the modern city, mall, and workplace.

Americans can see templespace in a contemporary setting in their capital, Washington, DC. Some of the finest buildings—the National Art Gallery, the Supreme Court Building, the National Archives, and the Lincoln Memorial—are "temples" to our secular democracy, and hallowed places for our citizens. Tourists from around the world come to visit them in endless streams.

Another dramatic example of templespace—this time a religious temple for worship—exists on the other coast of America: Robert

Schuller's Garden Grove, California, Crystal Cathedral, completed in 1980. The flamboyant evangelist raised over $18 million for the project, and hired Philip Johnson as the architect. Dedicated "To the Glory of Man for the Greater Glory of God," it attempts to be a New World Jerusalem, translucent and iridescent, with sunlight coming through 10,000 windows. The ninety-foot doors are open, as fountains shoot jets of water heavenward and thousands of worshipers are bathed in light. Over 2,500 years after the Greeks built the Parthenon in Athens, we have a modern updated California-style Parthenon, which, like the original Parthenon, makes use of clarity, light, and openness.

No more cavespace or pyramids for the Greeks or for those of us in the twentieth century who live and work in buildings dominated by glass and light. The Greeks brought us out of the dark—physically and metaphysically—and turned their love of wisdom into a new discipline: philosophy. They also laid the foundation for what has come to be called, centuries later, popular culture.

Forming tiny city-states, all independent and some democratic, they instituted systematic political science and were the first to write history as a subject distinct from myth and legend. Herodotus, long known as the "Father of History," was a "Father of Folklore" as well. He traveled far to find out all he could about the past. So did Thucydides, presenting history as a guide to the good life.

The epic poem, the drama, the lyric, the oration, and the dialogue all owe their origins to Greece. Each had its own forms and principles of composition, and all have continued to influence the world ever since the Periclean Age. In Aristotle, who lived in Athens from 384 to 322 B.C., we have the greatest codifier of all times; the father-founder of many sciences, whose influence on human thought has been continuous. So has it been in political theory, democracy, and popular culture.

Words rooted in the Greek dominate this book, and many aspects of our lives: *demos, ethnos, heros, theos, logos, eikons, mythos, symbolon*, for example. We are linked to Greek thought with unbreakable chains.

At the same time, in our Age of Multiculturalism, we must be aware of the continuous impact of non-Western sources, such as the Byzantine world, the Arabic world, the Near East, African, and

Asiatic. The Byzantine world, for example, represented the most direct link to the immemorial civilizations to the east. The Islamic world not only dominated much of North Africa, but spilled over into Spain. There is no boundary that can long contain or restrain the human spirit.

Within the time frame 700 B.C. to A.D. 700 great religious leaders—Confucius, Buddha, the major Jewish prophets, and Mohammed—flourished; in the middle of that period Jesus was born. They affected all aspects of world culture, and their "popularity" dominates human history.

Having said that, we return to the Mediterranean, and to another powerful civilization, which has directly shaped much of Western history, language, and politics; and beyond that, contemporary thought and action in the rest of the world. It is time to visit the Eternal City.

ENDNOTE

Temples predated Greece and now dot the world map. An early classic study is Seth Nola's *Cypress—Ancient Tombs and Temples* (New York, Harper and Brothers, 1878). Two excellent more recent books are Rex Martienssen, *The Idea of Space in Greek Architecture* (Berkeley, CA, University Press, 1964) and Vincent Scully, *The Earth, Temple Gods, and Greek Architecture* (New York, Praeger, 1969). A visit to Washington, DC, will show how we have Americanized the temple in our own time.

Forumspace

In 146 B.C. the Greeks were conquered by a new aggressive people—to the Greeks they were barbarians—known as Romans.

Though the Romans developed and kept their own language, Latin, they absorbed much from the artistic and intellectual culture of the Greeks. In the next three centuries they assembled an empire stretching over the then-known world west of Persia, including Egypt, Greece, Asia Minor, Syria, and much of what we now call Western Europe. The nerve center for this vast empire was the Roman Forum—perhaps the most historic and valuable small piece of real estate in the world. Just as the caves epitomized prehistoric humans, the pyramids the Egyptians, and the temples the Greeks, so did the Forum for the Romans.

How to approach the Roman Forum—and to explain what it has meant for so long to so many?

Here in the marshy gully between the Capitoline and Palatine Hills, some of history's greatest dramas were enacted. Here Romulus, Remus, Horatius, Caesar, Cicero, Augustus, and Trajan proclaimed *Roma caput mundit*—Rome, conqueror of the world. Just over there, Julius Caesar, who conquered so much but offended so many, was stabbed to death. On that rostrum Cicero's severed head was hung.

Rome, the Eternal City. The home of many of the ideas and achievements that make America proud to think of itself as the New World Rome, duplicated in part in our own capital, Washington. Rome gave so much to the world for so long that one hardly knows how to summarize its contributions in a few pages. Yet we must try, centering on the man destined to become the godfather of popular culture.

The Tiber River winds its way through Rome like a snake, as it has done for thousands of years; if it could talk, what tales it could tell! Rome of the Emperors, the Republic, the Caesars, the Popes.

Never had so many people been governed from a single city; never had such peace, the *pax Romana*, prevailed. *Roma, non basta una vita.* One lifetime isn't enough to see Rome or tell its magnificent story.

What then to do? Come with me and stand in the Forum, the very heart of Rome and all it stood for, and recall the life and contribution of a man, Marcus Tullius Cicero, who is central to the story we have set out to tell.

The word "forum," akin to Latin *foris* (outside) and *fores* (door), came to mean a marketplace or public place to conduct judicial and public business. There were forums, or open gathering places, in most Roman cities. In legal terms the word is equivalent to "court" or "jurisdiction." Basilicas, law courts, temples, and other public buildings clustered there. In short, the forum was the heart of the city.

Here the Senate met, policy was enacted, officers were elected, Caesar was assassinated, and triumphal processions were held on great occasions.

Like thousands over the centuries, I have gone to Rome's Forum and felt the power of the past speak through the majestic ruins. It is steeped in memory, glory, and blood. If popular culture has a leading shrine in the ancient world, this is it.

The process of Romanization carried with it much of the popular culture of the vast empire, and made Latin the universal language of the empire. The Forum was the center around which the known world, the *orbis terrarum*, turned.

The Forum, old heart of the Roman empire, is traversed by the *Via Sacra* (Sacred Way). In the twentieth century we can only imagine how splendid it was at its zenith. For centuries it was an unrivaled power center. Today it is crowded with the jumbled ruins of temples, arches, basilicas, and other buildings spanning over a thousand years.

Here the Vestal Virgins lived and tended to the Eternal Flame. The only man allowed in their temple was the Pontifex Maximus (Greatest Bridge Builder). That title was later adopted by the Popes, and the abbreviation *Pont Max* appears on many papal buildings.

The Roman Forum has seen hard days, being totally buried during the Middle Ages, to be excavated in the twentieth century by

Mussolini, anxious to portray himself as a new Caesar, *Il Duce*. He has gone, but the Forum still attracts millions.

The first-time visitor is apt to be disappointed by the sight of overturned boulders, weeds, and empty spaces. That even this remains is incredible, since it was used for years, like the Colosseum, as a quarry, and used later on as a *campo vaccino* (cow pasture).

Countless monumental places in Rome have captured the popular imagination—such as the Pantheon, home of all the gods, still intact and impressive after twenty centuries; the Colosseum, the ultimate sports arena in the ancient world; the Circus Maximus, where chariot races were run centuries ago—the likes of which reappear in modern films such as *Ben Hur*; and the Baths of Caracalla, in ruins now, still the center of spectacular summer grand operas. But none has such profound memories as the Forum.

We shall not attempt to retell the oft-told tale of that forum and city, but will focus on a pivotal person in our story—Marcus Tullius Cicero, born on January 3, 106 B.C., on the family homestead outside Arpinum, sixty miles southeast of Rome. Arpinum still boasts a fine statue of Cicero in the town square. I like to walk through the Porta dell'Arco, a gate in the old wall.

The local town school is named after Cicero, and from the town one can look out at the fertile, sun-drenched rural Italy that he loved. Who there could have guessed, then or now, that he would play a pivotal role in history, becoming the godfather of our popular culture? That his words would outlast the mighty Roman Empire?

One can spend a lifetime answering that question. We know more about Cicero than any ancient figure—perhaps more than any historic figure up to the Age of Print. Yet in some ways we hardly know him at all.

The best biography, by Gaston Boissier, was published in France in 1857, and translated into English forty years later. His words still electrify, for there is passion in his narrative. Cicero makes Boissier tingle: "What furious bursts of anger in his attacks!" Yet another French historian, using the same sources (Jean Carpicino) accuses Cicero of almost every failing and pettiness under the sun.

England's Anthony Trollope strongly defended Cicero against what he considered the shoddy treatment given him by writers such as Macaulay, Mommsen, and Froude. Whom then shall we follow?

Why not turn to Cicero himself and that great masterpiece, *On the Good Life*? It was written when Cicero had been driven into retirement, and all the glories of his earlier career had vanished. Never mind—he would take refuge in writing, there in his Tusculum villa, and try to make sense out of his life and world. The result was an astonishing burst of creative energy and creativity, equaled by few such moments in history. In twenty months he completed eight major works and a dozen minor ones. How did he do it? And how can those of us who now seek our Tusculum villa learn his secret?

He gives us strong hints. He did not depend on the latest trendy insights and spontaneous thoughts. He drew on all he himself had done and written over the years—editing, paraphrasing, revising, blending, summarizing. He always felt free to use what he wanted to when it suited his purposes.

Scores of classicists know more Latin and Greek than do I. I claim no new texts or discoveries, though I try to keep up with what others have discovered. What I want to do for Cicero is what he did (much better, you may be sure) for many others: bring the past to life in a fresh and convincing way.

A Herculean task, and I may well fail. But one thing I have already accomplished: Cicero has become my friend, and he has revivified the whole notion of how I might live my later years. His essay *On Friendship* has turned my life around, centuries after he and all his friends have died. It might do the same thing for you.

Instead of sound bites, he gave us sound learning. He went to Greece to study with Molo, who taught him to emphasize content and clarity. Before he was thirty, back in Italy, Marcus Tullius Cicero was widely regarded as the leading Roman orator. As a trial lawyer he had no peer. His philosophical writings included translations, dialogues, and discussions. His rhetorical works, centering on oratory, are still standards for the field. Conversation, said Cicero, should be easy and flexible, spiced with wit. Always be a good listener, respect your opponent, and leave graciously. Politicians of the world, take heed!

Passionately devoted to the Republic, Cicero was neither a populist nor demagogue. He knew that catering to deep and inherent prejudices might unleash the mob, and ignite a political tinderbox.

But he deplored the easy solution—buying off the people with bread and circuses. Not charity but prudence should guide the leaders. Yet that prudence must always be tempered with compassion. No government can endure except with the consent of the governed.

Ours is a government of laws, not individuals. If we can locate the principles and the integrity, we may even be able to find someone who exemplifies them. Celebrities need not apply—people who are famous for being famous. Cicero would not have approved of any such candidate—and neither should we. Rather than rising above the worst aspects of mass culture, many public officials and candidates have become a part of it. They thrive on the glitter of tinsel, shadows, moral evasion, and phony fellowship. People are tired of a system that glorifies cotton candy when we crave and need real food. Where, in fact, is the beef?

Cicero, being a good historian, would remind us (as would his candidate of choice) that democracies and republics are often divided. Over 130 years ago a civil war rent our nation asunder and almost destroyed us. A century before that a raggle-taggle army made up mostly of farmers staked their lives and sacred honor (a word most dear to Cicero) on revolution and, following a man who might seem to Cicero a latter-day Roman (George Washington), we became a republic. Well done, Cicero would have said. Farmers make good citizens, soldiers, and presidents.

The love of farming and the landscape shaped his work, images, and ethics. (Here again—note both the Washington and Jefferson parallel.) The bank in which true pleasures are kept, Cicero wrote, is the earth itself; it honors every draft and invariably returns the principal with interest. A well-kept farm is the most useful thing in the world—and the best to look at. From farming his Founding Fathers and ours got "an unbelievable amount of enjoyment."

To love the land—the smell of the soil—does not mean one has to slight the universal. We are all bound together by a common chain, Cicero said; anyone alive in the 1990s knows how true this is. Human unity, sympathy, community—these are golden threads running through Cicero's writings. A latter-day Cicero would search for new global myths, in sources Cicero never knew about—film, television, radio, mass print. He would observe their work on various levels, simultaneously. Cicero, more than anyone else, would urge us not to dig in the dust of Rome just to find archaic myth and truth. We should look to him, and to Rome, for inspiration, insight, and solutions that will allow us to survive in the present, and have a future to plan and anticipate.

Cicero was one of history's most prolific letter writers, and over 800 of them have endured. The letters to his friend Pomponius Atticus, whom Cicero called "my second self," are a treasure trove. Easily deceived and often inconsistent, Cicero admitted in one letter: "I know I have been a regular ass." How many of us are willing to say as much?

For centuries men and women have wondered what it would have been like to sit and talk with Cicero. One of these is the fictional character Mrs. Blimper, in Charles Dickens' *Dombey and Son* (1848):

> If I could have known Cicero, and been his friend, and talked with him at Tusculum, I could have died contented.

Latin was Cicero's road to immortality. For centuries it was a minor branch of the Indo-European language family, confined to a small coastal area of western Italy. About 400 B.C. it began to spread across the peninsula, evolving from a crude dialect to a powerful instrument—eventually to become the glue that held a great civilization together.

Latin drew heavily on Greek. Cicero relished Greek models and ideas, and studied them carefully and successfully. Along with poets such as Lucretius and Catullus, Cicero nurtured and shaped Latin into a sensitive means of expression, reflecting both meanings and values. He knew that speaking and writing are not isolated acts, but part of a whole social process—the basis of true community. This was one of his major insights.

What he and other writers did was to distill and transmit Greek learning into Latin. They added devices of their own: emphasis, contrast, repetition, rhetorical questions, exclamations. They learned how to subordinate clauses and tenses to get new and subtle meanings. They created cogent compact prose and gave it a harmony that was almost musical. They established Latinity.

That is why the Roman Catholic Church has clung to Latin for many centuries—as have scholars, scientists, doctors, and bureaucrats everywhere. That is why our flowers have Latin names, and why our medical prescriptions are in Latin. That is why our whole legal profession still functions on Latin phrases and concepts.

Many great thinkers and writers of the past revered Cicero. Longinus compared him to a spreading great fire, leaping and ranging over whole fields. Think of the words Cicero either coined or popularized: infinity, image, instance, notion, comprehension, to name a few. But he wanted to do much more than shape words. He wanted to shape the greatest structure of government and empire the world has known. He believed he could extract and express universal laws, goals, and concepts that would serve not his age, but any age. Most of all he believed in a harmony of all being—what the Greeks liked to call the music of the spheres. With such a grand goal, he did not hesitate to draw from the various sources in such measure and manner as suited his purpose. He did just that. From the Greeks he learned about Egypt, Assyria, and Persia. He was an avid student all his life. Four Greek writers were the pillars on which he built: Pythagoras, Plato, Epicurus, and Zeno. He greatly admired many other writers: Panaetius, Posidonius, Arcesilaus, Carneades, and Mintiochus, for example. He was proudly eclectic, trying to combine a manual of civic virtue with a system of applied ethics. Cicero borrowed from many, and many have borrowed from him. One was John F. Kennedy, who capped his 1960 inaugural address with this vintage Cicero: "Ask not what your country can do for you; ask what you can do for your country."

Cicero has gone, and Kennedy too, but the thought they shared so dramatically remains. If more people acted on this thought we might experience a great political and social reformation. The dead still have tales to tell, if we have ears to hear. If we listen carefully, we may realize that the past is prologue, and that there is nothing new under the sun.

Few writers have left so rich a legacy as did Cicero. His philosophical writings included translations, dialogues, and discussions. His rhetorical works centered on oratory, methods of drawing conclusions, and case studies of famous orators. Then there are the hundreds of letters. Cicero may be the greatest letter writer of all time. Consequently we know Cicero as a whole man—a complex individual from ancient times who speaks to us today.

What makes Cicero, dead for so long, still seem so very contemporary? Two reasons. He was a synthesizer and a transmitter—two words that also sum up our Electronic Revolution. He was a master at paraphrasing, summarizing, analyzing—he would have been at home with sound bites, thirty-second spots, and the ins and outs of our electronic campaigning. How he would have responded to all the goings-on in the Clinton White House! Alas no one of his stature is doing that now.

Cicero ranks first among the world's popularizers, passing on to Rome and then to the whole world many treasures of older cultures. He knew that being simple and straightforward is never simple; that being widely understood and enjoyed is the greatest thing any writer or thinker can achieve. He himself enjoyed the ephemeral but sought the perennial. Cicero was a true humanist, in the best tradition of that word, shown by his many writings, orations, and letters.

* * *

Thomas Jefferson, another great humanist and letter writer, said that the tree of liberty must be watered with the blood of patriots. Cicero's blood was shed to that end. For delivering his powerful *Philippics* against Mark Antony, Cicero was proscribed—named a public outlaw and traitor who could be killed without penalty.

Antony's hired killers tracked him down and caught him at Formiae on December 7, 43 B.C. For a few minutes he gazed keenly at them—he had been reading Euripides' *Medea* when they entered—leaning his chin on his left hand. Then he calmly bared his throat to their swords. Cicero was sixty-three.

His head and hands were hacked off and presented to Mark Antony, who ordered them nailed to the speaker's platform in the Roman Forum as a warning to those who opposed him. One last indignity: Antony's wife came to the Forum and stuck a pin through Cicero's tongue.

The Romans shuddered. "They believed they saw there," the historian Plutarch wrote later, "not the face of Cicero but the image of Antony's own soul."

His tongue has never been silenced.

ENDNOTE

From the vast library about Rome I chose my two favorite books: J. P. V. Balsdon, *Roman Civilization* (London, Penguin, 1969) and R. M. Ogilvie, *Roman Literature and Society* (New York, Viking Penguin, 1980).

Among contemporary Cicero scholars and interpreters two stand out: D. R. Shackleton Bailey and Michael Grant. Bailey's masterpiece is *Cicero: A Biography* (London, Duckworth, 1971). Among Grant's many books and anthologies, I strongly recommend *Roman Literature* (Harmondsworth, U.K., Penguin, 1964) and *Cicero: The Good Life* (Harmondsworth, U.K., Penguin, 1971). Especially helpful to me was J. C. Rolfe, *Cicero and His Influence* (London, Harrap, 1923).

Medieval

The terms "medieval" or "Middle Ages" are flexible and elusive terms—as is our current favorite, "postmodern." In the "middle" of what? And for how long? Were they "dark" at the beginning and "enlightened" at the end? Who decides? Was this the "Age of Faith"? Faith by whom—in what?

There is more agreement on the three dominant aspects of medieval popular culture—religion, hierarchy, and combat. The monk, the king, and the knight held sway. Gothic cathedrals sprang up throughout Europe, and are still the glory and high achievement of Western civilization.

The Roman Catholic Church was at its peak of power, and the crusades were a demonstration of religious fervor and devotion. The medieval Church was one church and all churches. Everyone belonged to the Church, and the Church belonged to everyone. It was the glue that somehow held society together.

Just as we speak of a black hole in the universe, we speak of the Dark Ages in civilization. Though poorly defined, the term is widely used. I recall a definition one of my students wrote on his examination: "A long period when the sun didn't shine."

We must do better than that. We know that five hundred years ago, well-meaning scholars chose to dismiss much that went on between the decline of the Roman Empire and their own time as dark and barbarous. Sometimes they used the term "Early Middle Ages" as well. In this "middle," we often cover a thousand years, between the fifth and fifteenth centuries A.D. Some middle!

There are no complete breaks in history—no clear-cut beginnings, middles, and ends. The more we try to dismiss the past, the more it pops up in the present. We should avoid glib labels that deceive us of the content—cutting up history into decades, periods, and centuries is essentially arbitrary. Once we plant the idea of the Dark Ages or the Middle Ages in our heads, we become its prisoners.

Instead of speaking of the Dark Ages, we might use another metaphor—the Frozen World. The glory that was Greece and the grandeur that was Rome combined to make the classical world, giving us the foundation of Western civilization. As Kenneth Clark points out in the TV series *Civilization* (1969), after the fall of Rome in the fifth century A.D., nearly all the life-giving human activities he labels "civilization" had been destroyed or forgotten in Western Europe. The Big Freeze set in.

Those determined to preserve the classical heritage fled to inaccessible fringes of the known world: Europe was dominated by pagan wanderers and barbarians. Only when Charlemagne was crowned Holy Roman Emperor in 800 A.D. was contact with ancient cultures reestablished.

Like the anonymous cave painters, Menes, Tutankhamen, Pericles, Plato, Cicero, and Caesar, Charlemagne reaches a high plateau in the history of popular culture. By combining the church and the empire—the cross and the sword—he restored peace and order, and thawed out the Frozen World. He spent almost half his life in battle. He boldly crossed the Pyrenees and brought northeastern Spain, occupied by the Arabs, back to Christianity; overthrew the barbarian kings in Italy, struck deeply into eastern Europe, and (except for England and Ireland) created an empire like that of the Latin Christian world. The center of power had moved from Rome to Aachen.

Now scholarship and learning would be revived. Though he understood Latin, Charlemagne could barely read and never learned to write. Yet he did all he could to revive ancient languages, copy disintegrating manuscripts, and bring to his palace scholars from England, France, Italy, and Spain. He fostered commerce and created a new and more reliable coinage, based on silver. (A pound of silver was divided into 20 *solidi* and 240 pennies.) Centuries later, America's ever-popular Benjamin Franklin would write in *Poor Richard's Almanac*: "A penny saved is a penny earned." In a real sense, the civilization Franklin inherited was saved in part by Charlemagne.

Yet the long shadow of the "Dark Ages" hung over Europe, and much of Charlemagne's empire crumbled. Hordes of barbarians, Vikings, and Norsemen attacked Western Christendom. Thousands were killed, robbed, or carried off into slavery.

One reason leading humanists such as François Rabelais (1494-1553) could speak of the "Dark Ages" is that many centuries were for him—and for us still are—shrouded in darkness and uncertainty. Things literally fell apart—cities, roads, trade, education, uniform laws—all that had brought high civility to the Western world. Putting them back together was slow, arduous, and painful. Much of the process is still shrouded in darkness. There was no one European model, no strong center, no clear national borders. There was only the Christian Church, with its many problems, and the term often used for what is now Europe: Christendom.

No one expressed this dilemma better than Bishop Augustine of Hippo, in North Africa—later to become Saint Augustine. He wrote even as the barbarians were at his gates and the whole Roman Empire was crumbling. To him the world was ending—the apocalypse had come. Humankind could only repent, pray, and wait.

In fact, the old Roman order and ideals did tend to fall apart. Whole provinces were looted and slaughtered. Rome itself was sacked, chaos prevailed. The times were indeed dark.

* * *

On Tuesday, November 27, 1095, Pope Urban II held a public meeting at Clermont to make a great announcement. The bishops of France and neighboring lands came, along with thousands of clerics and laypeople. The crowd was too large for the cathedral, so the papal throne was set up in an open field. Eastern Christendom had appealed for help; the Turks were advancing into the heart of Christian lands. Western Christians must march to rescue the East! Rich and poor must go, to do the work of God, who would lead them. Everyone should be ready to leave home by the Feast of the Assumption.

The Pope was a great orator and the response was tremendous. "*Deus le volt!*"—"God wills it"—shouted the crowd. The groundwork had been laid for a series of crusades—one of the great popular movements in Western history.

We need not recount that extraordinary saga here, but should point out that medieval people were convinced that the Second Coming was at hand, and they must go out to do good. The Church taught that sin could be expiated by pilgrimage; the Holy Land must be recovered before Christ came again.

Some idea of the climate of popular opinion that fostered a series of crusades emerges from the Children's Crusade. In May, 1212, a twelve-year-old shepherd boy called Nicholas brought to King Philip of France a letter which, he said, had been given to him by Christ in person. Nicholas was to lead a crusade. The king, unimpressed, told him to go home.

Instead, Nicholas went around France urging children to rescue Christendom. The seas would dry up before them so they would pass, like Moses through the Red Sea, safe to the Holy Land.

Thousands of children massed at Vendôme, all under eighteen years of age. The youths started out on foot, though Nicholas rode in a gaily decorated cart. Locks of his hair were collected as sacred relics. En route to Marseille some died, others deserted. But a sturdy band got to the port and waited for the sea to divide before them.

It didn't. The children waited for God to relent. He didn't. Then two merchants, Hugh the Iron and William the Pig, offered to carry the children free of charge to Palestine. They rejoiced and set out to sea in seven vessels. There was no further news of them for eighteen years. Then it was reported that they had all been slaughtered.

* * *

After many dreary years things improved. Even though recent investigation has added new insights—some historians now speak of a twelfth-century renaissance—material on these "dark" centuries is scant, and often, so far as the illiterate peasants of popular culture are concerned, nonexistent. Work, hardships, sudden death, famine, disease, plagues, wars, feuds, sacks, and robberies took their toll. The average life expectancy fluctuated between twenty-five and thirty. Infant mortality was as high as 40 percent. Death was a constant visitor.

Life was determined largely by nature—sun, rain, and snow, seasonal rhythm, day and night. People were at the mercy of forces they neither understood nor controlled. There was no exact way to measure time. Work began at sunrise with the cock's crow and ended at sunset. Living quarters for peasants usually consisted of a single room, serving as bedroom and workroom, with an open hearth. Floors consisted of tamped-down dirt; "windows" were openings covered by shutters. Frequently animals lived under the same roof.

Epidemics and diseases followed famines. Medical skills were negligible; plagues were devastating. Sanitary conditions were deplorable even among monks and nobles. Strict monastic rules limited full baths to twice a year—Christmas and Pentecost. Religion and superstition prevailed. The family was the smallest and most important sphere of life, characterized by a strong sense of unity. The husband was the head and chief authority. Everyday life consisted of work, meals, and occasional entertainment—all dependent of class and social status. The overarching rule: a place for everything, and everything in its place.

In the modern popular imagination the Middle Ages fostered sumptuous castles, homes of noble lords and knights who went about doing good deeds. Actually castles were confining and crowded, and the "palaces of princes" hardly palatial by our standards. Castles were built largely for defense, and were cold, dark, and forbidding. The castle became an instrument of power in the hands of the warrior class.

What education there was, and what preservation of the fractured past, was reserved largely for the clergy and monks, and took place mainly in monasteries. They became centers for the veneration of saints and miraculous activities, which were important aspects of popular culture and belief. Monasticism was a highly respected way of life; monks prayed for everyone, and everyone supported them. They were largely responsible for preserving what we know as our "classical heritage."

The model for Western monasticism was first set by Saint Benedict of Nursia (c. 480-547), centered in monastic quarters in Italy. Many bishops, even popes, came from the ranks of monks, who assumed great importance throughout Europe. Monasteries pioneered agricultural reforms, providing for agricultural surpluses and fair land acquisitions. This was especially true of the Cistercians. By 1153 there were 344 Cistercian communities in operation.

Medieval culture was basically agrarian, and popular culture centered on the peasantry. The manorial system was tied to the soil or landed property. Most peasants or serfs lived and died under the manorial system, although some of them were free.

What did the peasant do? He tilled the soil and raised livestock; used the plow; cultivated the soil with his own hands. His work was spread out over the entire year, determined by nature and the weather. In addition to house chores women helped in the field during "high season," and sometimes assisted as oxen drivers during plowing. They milked cows, sheep, and goats, and made textiles. But there were things peasants could *not* do. According to the *Imperial Chronicle* of the mid-twelfth century, peasants could not wear any colors other than black or gray. Seven yards of linen for shirts and pants was a top limit. A peasant possessing a sword would be bound and led to the church fence, tied, and thoroughly beaten. These same kinds of limitations would apply to slaves many centuries later.

Scholars now know something about the work life and habits of peasants and serfs, which reflect many habits of their culture. But we have almost no information about the rest of the everyday domestic life of the peasants.

Some of our best insights into faith, and how it affected and directed ordinary people, come from literature, liturgical plays, and folklore. By common consent two literary giants of the Middle Ages are Geoffrey Chaucer in England, and Dante Alighieri in Italy. In his *Canterbury Tales*, Chaucer places thirty-one pilgrims—most of whom would have been considered "lower class" or "in trade" by aristocrats—on a summer pilgrimage. The travelers include a plowman, a weaver, an innkeeper, a reeve, a seller of church pardons, and the bawdy, five-times-widowed Wife of Bath. Chaucer has sought out those of lower rank, and has "spoken with them everyone." He is a medieval exponent of popular culture.

The Canterbury Tales gives us an intimate view of life in a feudal-commercial society, at a moment when life is stable and enjoyable for many people. Chaucer finds his pilgrims amusing and in many ways admirable. We wish we had been along with this band of pilgrims, sharing their laughter and their tales.

Dante, on the other hand, shows us an Italy caught up in a bloody power struggle—and he is on the losing side. Charles of Valois invaded Florence and crushed Dante's faction in 1301. Dante himself was exiled, spending the rest of his life roaming from one Italian city-state to another, brooding and condemning his enemies. He is buried not in his native Florence, but in Ravenna. He never came home.

Dante, a great poet, did more than condemn his enemies: he invented a hell, and put them there. His book, *The Divine Comedy*, makes universal some of the deep feelings and conflicts of his troubled times. Full of intense suffering and searching, it probes the depths of the human soul. From the dark caverns of his exile he confronts the problem of evil head-on, and speaks to every age, not merely his own. *The Divine Comedy* has special power and poignancy for the twentieth century.

How different Chaucer and Dante are—yet they are both spokesmen for humanity and the people. They are local and universal at the same time.

We can say the same for Thomas Aquinas (1225-1274), whose *Summa Theologiae* surveyed all of knowledge, demonstrating that faith and reason need not be in conflict. His hierarchic view of the universe and of society provided the foundation of official Roman Catholic philosophy, and is firmly in place today.

The many regular festive occasions were apparently celebrated fervently, such as the feasts of St. Nicholas and St. Silvester, and the anniversary of the village church's consecration. Baptisms and weddings were special events with singing, dancing, and drinking. Drunkenness was common, but leisure and free time were rare. The idea of entertainment, as we understand it today, hardly existed.

How "dark" were the Dark Ages? There is no simple answer. These bleak and difficult centuries were multifaceted. Many people had pleasures and privileges not available to ordinary people. Daily life in the monasteries, manor houses, princely courts, and cities allowed for a wide range of options. Everyday life existed within definite bonded relationships. Even the everyday life of monks varied widely among different orders and monasteries. Authority-based relations held wide sway. The authority of the Church was all-encompassing.

Surely the rosy views of medieval life reflected in modern musicals such as *Camelot* have little historical validity. Life was not free; one was born into social status, and few could change it. But how can we measure happiness, community, and hope over these long and little-known epochs? What can we say of their popular culture?

A particular "world view" dominated medieval European culture. Life on earth was not free, but was punishment for original sin. The only salvation lay through the Church, and Christ. This life was a preparation for the afterlife—heaven or hell. Peasants performed compulsory labor. For the most part, people resigned themselves to their fate.

If there were "dark times," who or what turned on the light? The Renaissance? The Reformation? Changing weather patterns? Improved technology? These questions are perpetually debated. Students of popular culture like to speak of "The Big Shift," which occurred in what we call "modern times." It was then that two traditions—the great tradition and the little tradition—took definite shape, and popular culture took on a life and meaning of its own. Our next task is to define those two traditions, and the "modernity" that followed.

ENDNOTE

There are many excellent summaries of the Middle Ages. The famous eight-volume *Cambridge Medieval History*, published in England by Cambridge University Press from 1911 to 1936, was reissued in a two-volume summary in 1952, edited by C. W. Previte-Orton. Other excellent sources include James Thompson, *A History of the Middle Ages, 300-1500* (New York, W. W. Norton, 1931); Eileen Power, *Medieval People* (New York, Barnes & Noble, 1963); G. G. Coulton, *Medieval Panorama* (New York, Meridian, 1955); and F. B. Artz, *The Mind of the Middle Ages* (Chicago, University of Chicago Press, 1980).

The American Historical Association Service Center in Washington, DC, has issued various bibliographical essays on work dealing with the Middle Ages.

Modern

When we sail on the vast and choppy sea of "Modern," we have the same problem that is met whenever we divide history into pieces and apply labels (such as ancient, classical, medieval, modern). The divisions exist in our minds, not in the real world.

"Modern" is especially elusive. When do modern times begin? What makes them modern, and where? The word's derivation (from Latin *modo*, just now) reminds us that "now" is a moving target, soon to become "then." Modern is a period, the dictionary says, "extending from a relevant remote past to the present time." All the past is relevant—not nearly as remote as some say. Do we talk about the "immediate" past? To be modern is to be new—is there anything new under the sun?

Let's leave this semantic jungle, and concentrate on one person who had two significant titles: Father of Modern Science and Father of Modern Philosophy. René Descartes (1596-1650) provides a good starting point for an essay on modernity.

On the night of November 10, 1619, the Jesuit-trained French youth Descartes had three dreams. The Angel of Truth visited him, and revealed a secret that laid the foundation of modern science. The time had come to rethink human knowledge, to begin *de novo* (from the beginning). Tradition, superstition, and papal authority must be left behind. A new method was needed—one that was rational, systematic, clear, and distinct. The secret: even a mediocre person can achieve great things with the proper method. And you, the Angel told Descartes, will supply it.

He did. His astonishing analytical genius, including the invention of coordinate geometry and works on methodology and metaphysics, was demonstrated in his *Discourse on the Method of Properly Guiding the Reason in the Search for Truth in the Sciences* (1637). His famous dictum, *Cogito, ergo sum* (I think, therefore I am) helped set us on the rational, empirical, mathematical course that underlies the modern scientific and technological world—and today's popular culture.

Was the modern world born with Descartes in the seventeenth century? Some set the date much earlier. Others prefer to use terms such as Renaissance, Industrial Revolution, and Enlightenment for the sixteenth, seventeenth, and eighteenth centuries. Just what *is* "modern"?

Another question springs to mind. Suppose Descartes had questioned the Angel, asking, "Why did you leave out the intuitive?" Rational thought and mathematics are fine—but what about all the unconscious sources, the sudden flashes, the dreams? Isn't that where the master ideas come from? Can I not draw from the deep enduring wisdom of fairy tales, fables, myths, legends?

Descartes didn't ask the question—we can never know what the reply might have been, or how the modern world might have been different.

What we *do* know, and have to factor into twentieth-century thinking, is that fairy tales, myths, and legends live deep in the human psyche: primitive and primordial. They are among our first and best teachers. We learn from them. We fear wolves (one ate little Red Riding Hood's grandmother), hungry giants (Jack confronted one at the top of the beanstalk), and the werewolves that howl when the moon is full. The modern world was changed forever by the later work of Sigmund Freud and the science he invented—psychoanalysis. He has taken us into modern worlds that Descartes never spoke of—or never dreamed of. Has Freud done more to shape popular culture than René Descartes? Who can say?

* * *

"Modern," like other words we have employed, is a sticky concept, and invites confusion. If we are not sure where modern begins, we know it has had a wide and diverse application, and means many different things to many different people.

When did modern times begin? From the Euro-American perspective, when the scientific revolution took root in the West? When European countries began to industrialize? To urbanize? Or when? A key word for modern thinking is "progress"—a myth or a reality? Are we getting closer to perfection, or closer not just to tragedy, but a new tragedy? Values have been altered—people began to think and act in new ways. Do these changes cross national boundaries? Do they go through set changes? How do the ordinary people fare in all these changes?

Can we say the modern world is moving in one clear direction or another—for better or for worse? Or (as some argue) have we left the modern world and moved into postmodernism?

There are many questions, few clear answers. As perspectives, leaders, and trends change, so do reactions to modernism, industrialism, and urbanism.

Modernism implies a self-conscious break with the past—a determination to be up to date, and to search for new ideas and forms. These ideas involve such broad categories as taste, style, mind-sets, and attitudes. They in turn affect many of the more tangible things in our lives—our clothing, speech patterns, popular arts, and entertainment. This is the driving force behind our popular culture, largely centered in the twentieth century.

Instead of trying to summarize our incredible century, in which all of my life (and most of my readers') is centered, I shall use the mosaic idea put forward earlier, and present several tesserae, or pieces, that illuminate our times.

The first three pieces interlock and overlap. They are people who became not only famous, but institutions known around the globe—Disney, Ford, and Edison. They gave us Mickey Mouse, the Model T, and the electric lightbulb. Long dead, they are among the three best-known people on the planet. In the political realm, we shall salute F.D.R., who gave us the New Deal during the 1930s, and led our country during the century's cataclysmic struggle, World War II.

Then we shall look at how this modern twentieth century changed our popular architecture, music, and dance. Finally, we shall ask: Has the modern age ended? Have we moved into the postmodern?

Let's begin at what might seem an unlikely place—with Mickey Mouse. Mickey and I are about the same age; but as I have grown older, he grows ever bolder. Once a three-fingered character in animated cartoons, he has become the host in Magic Kingdoms on three continents. He presides over 600 stores, moves freely in all world markets, and is known by dozens of different names, such as Michele Souris, Michele Topolino, Miguel Gatonocito, Miki Kuchi, Mugge Pigg, Mikki Hirri—and who knows how many more in India's sixteen official languages?

Mickey leaped out of the film into his own three-dimensional world—Disney World. You can see, talk to, and touch Mickey. He heads a multimillion-dollar enterprise, in his own Magic Kingdom. He is a true icon of popular culture.

Mickey (like Athena of old) sprang full grown from the brain of a creator (Zeus for Athena, Walt Disney for Mickey). Born in Chicago in 1901, Disney drove an ambulance in World War I, returned to study in an art school, and drew illustrations for farm magazines. But his real interest was in creating animated animal cartoons. Unable to get money for his experimentation, Disney decided to try his luck in Hollywood. En route a mouse was born.

Going west on a train out of Chicago, he began according to his own account, to sketch mice. In what he liked to call a happy creative delirium, he gave one impromptu sketch a name: Mickey Mouse.

That delirium continued as Walt (to quote from the official version, which lapsed into a football metaphor) "developed into one of the most spectacular broken-field runners in the history of small industry" (*Saturday Evening Post,* October 31, 1953). To illustrate the point, *Life* magazine, on October 16, 1953, carried a picture of Walt's brother Roy, now the company's president, sitting proudly among some 3,000 of the 3,500 items which by that time bore Disney trademarks.

The early Mickey was functional, with black dots for eyes, pencil legs, three fingers per hand, a string-bean body, and a jerky walk. He talked for the first time in a 1928 cartoon called "Steamboat Willie," in which Mickey growled, squeaked, and amused. Who supplied the voice? Walt Disney. In fact, Mickey looked a lot like Walt.

In the 1930s, Mickey left the drawing boards and invaded the nation's homes and bedrooms. By 1940 over 2,000 companies were affiliated with Disney. The Ingersoll Watch Company sent its five millionth Mickey Mouse wristwatch to Walt as a special present. The watch would have a nostalgic revival in the 1990s. In the years between, millions of children everywhere thrived on Mickey's image. They wore Mickey clothes, bibs, and jewelry; read Mickey comics; played with Mickey toys, balls, and games. At the movies, they saw Mickey Mouse cartoons. At night they put on Mickey pajamas and tumbled into Mickey beds, to dream of a happy land where cats do not break through and steal.

Mickey's very existence shows how technology can create heroes independent of biology. Although Mickey is machine-made, he often confronts mechanical enemies. Sometimes vacuum cleaners or cement mixers set out to destroy Mickey, but to no avail. Tiny, agile, and compassionate, Mickey releases us from the fear of the very technology and impersonality that makes him possible. He achieves an illusion of independence from the real world. Yet he helps to shape it. Intelligence officers planning the Normandy invasion in World War II quickly agreed that the appropriate code name for the operation was "Mickey Mouse." It was tribute enough to turn the average rodent's head. But then, Mickey is no average rodent.

Nor was Walt Disney the totally creative innovator he has become in our mythology. His great inspiration grew from an earlier, much-admired model. Mickey's inspiration was the Model T.

* * *

Henry Ford believed that history was bunk, but nostalgia was golden. He also wanted to put America on wheels—and he did. He introduced his famous Model T Ford in 1908, a car for the multitudes. By developing the moving assembly line in 1910, he achieved phenomenal production results. By 1914, his workers completely assembled an automobile in ninety-three minutes, compared with twelve hours in 1913. By 1925, he could produce a new car every ten seconds. By 1927—when Mickey was ready to captivate Hollywood, and then the whole world—Ford had sold over fifteen million Model Ts, for as little as $265 each. He not only put America on wheels—he expanded popular culture and helped create a new machine-age folklore. The Model T became a member of the family, a close friend, with names such as Leaping Lizzie, Little Bo-Creep, Rolls Rough, Bouncing Betty, Lizzie of the Valley, or Passion Pot.

Ford jokes abound. They told of the old lady who sent old tomato cans to Detroit or the farmer who mailed his tin roof. They both got a Model T back by return mail. Had any king or magician ever turned tin cans into Fords? Another story told of the rich eastern banker who paid $15,000 for a fancy European car, but kept a Model T in the trunk to pull him out of the mud.

When popular culture and folklore overlap, incredible things happen. By 1923 Ford was making over 7,000 cars a day—and not spending a penny on advertising. After the Model T came the Model A and the V-8 engine. By 1940, one out of every seven American workers was in the motor car or allied business.

Henry Ford's life stretched from Gettysburg all the way to Hiroshima. As much as any American, he symbolized the vast technological changes that took place in those years.

If he ushered in the future, he doted on the "glorious past," amassing the largest collection of Americana then in existence. He not only wanted to preserve the American past—he wanted to own it. He especially wanted to own everything connected with his idol, Thomas Alva Edison. He moved Edison's entire estate, including the soil it was built on, from Menlo Park, New Jersey to his private domain at Greenfield Village, Michigan. He even exhibited a special bottle holding Edison's last breath.

Walt Disney, a frequent visitor to Greenfield Village after 1940, decided to outclass the Motor King. He built Disneyland in 1955, and envisioned a Disney World, built after his death in 1966, which took nostalgia to a much higher level. There were whole theme parks inside the compound—Main Street, Fantasyland, Frontierland, and Adventureland. Here is history devoid of problems, and devoid of substance. But his greatest tribute to technology, though posthumous, would come later when EPCOT (Experimental Prototype Community of Tomorrow) opened in Disney World. It attempted to turn fantasy into reality.

So did the Hall of Presidents, with Abraham Lincoln the dominant central figure. Programmed for sixty-five different body movements and endowed with electronic speech, the Lincoln robot delivers a sermon. The other robots listen dutifully and nod their agreement. It is history without controversy or complexity.

The operative word for this worldview, as put forth by Ford, Disney, and many others, is "progress," resting on science. It streamlines the earlier programs of the Industrial Revolution, as heralded by Condorcet and Turgot in France, Paine and Priestly in England, Jefferson and Franklin in America.

Thomas Edison (1847-1931), the "Wizard of Menlo Park," so admired not only by Ford and Disney but much of the world, was indeed a remarkable man. In a sense he invented inventing, building at Menlo

Park America's first professional research center and taking out 1,033 patents up to April 1928. His achievement has never been equaled.

Edison hired a whole staff to assist in multiple experiments. In 1869 he took out his first patent—for an electrical vote recorder. In 1877 he patented a "phonograph or speaking machine," from which many later phonograph and recording systems sprang. In 1879 he invented a successful incandescent lamp. The list goes on and on, year after year. He earned the title "Wizard" and a special place in the popular culture pantheon.

Disney, Ford, Edison—not merely three names, but three self-made men and pragmatic philosophers who gave America its three-fold formula for shaping the modern age—mechanization, mass production, and mobility. They turned these abstractions into available objects—the things on which our modern way of life rests. At the century's end, Disney World is a global theme park. Michael Eisner, Disney's current successor at Disney World, has moved into multiple activities and built a world empire. Ford Motors is still a leading car manufacturer, with new inventions and technology its major driving force. Con Edison, lighting up New York, is a leading power supplier.

For many the words "modern" and "American" have become synonymous. Our chief twentieth-century rival ideology, communism, has failed in most parts of the world, and people eagerly accepted consumer capitalism. We have become the world's leading economy and superpower. For many people, the American Dream is becoming a reality. They sit back, relax, see the stock market soar, and watch the Fords go by.

Many enjoying the prosperous times can remember grim ones. The Great Depression of the 1930s almost brought America to its knees, leaving a third of the nation ill-fed, ill-clad, and unemployed. Only the power and persuasion of President Franklin D. Roosevelt—who became the people's F.D.R.—pulled us through. His New Deal gave us new life and hope. But then we were thrust into a world war for which we were ill-prepared. Once again, F.D.R. led the free world, joining Washington, Jefferson, and Lincoln as a great leader. Yet it was over half a century later before there was a fitting Roosevelt Memorial in Washington. We will go there now.

F.D.R.

The moving finger writes—but at its own pace. Over fifty years after the death of Franklin Delano Roosevelt, the greatest president of our century—perhaps of our history—he has a fitting memorial in the national capitol.

For half a century those who went there to see his memorial found only a small table-sized slab of marble in front of the Archives Building, with his name and his own words noting that this is all he wanted.

Now F.D.R. has got what *we* want—and what he deserves: a seven-and-a-half acre garden memorial in Washington's Tidal Basin, putting him in the company of his true colleagues—Washington, Jefferson, and Lincoln.

There was a massive outpouring of praise and sentiment when the memorial was officially opened on May 2, 1997. It is easy to know why.

The monument itself breaks new ground. It is not a marble *object*, like the Washington Monument or the Lincoln Memorial, but a series of settings, all involving water. Roosevelt's history was, in a sense, written on water. He began his political career as Assistant Secretary of the Navy, recognized polio symptoms after a long and chilling swim, found relief for many years by bathing at Warm Springs, Georgia, and created the mightiest navy in history to play a key role in winning World War II.

His memorial has other things people remember and love: depictions of his wife Eleanor (perhaps the greatest woman of our century), unemployed Americans braving the Great Depression, and even his dog, Fala. We can walk into this memorial, and walk into his life. Many of us can also relive part of our own lives as well.

At this beautiful spot, in the reawakening of nature and memories in May, four images flash through my mind, so clear I can believe they are today's realities.

I am huddled around our Philco radio with my parents and grandparents, listening to one of President Roosevelt's Fireside Chats. They cling to every word and nod their strong approval. I am too young to understand just what the issues are. But I know that my family and friends, living in a Roanoke that is laying off workers and shutting down, take hope from these chats. Something is terribly wrong. "Don't worry," Dad says, "President Roosevelt is going to fix things." He was right. Has ever an American voice turned the nation around as did the voice coming over our radios?

Next I am standing outside the Norfolk and Western Railway station in downtown Roanoke, proudly bedecked in my new Boy Scout uniform, awaiting the arrival of President Roosevelt, who has come by train to open the new Veteran's Hospital in Salem. There he is, getting in his open touring car with that incredible, unforgettable, radiant smile. Since I am in the front of the crowd, I can almost touch him—and never forget him. Afterward, a dozen scouts from Troop 17 say just what I say: "This wonderful man paused for a brief second and smiled at me."

The years pass. Now I am wearing another uniform—on a Navy ship towing a damaged Polish tanker in the North Atlantic. The captain comes on the bridge with a strange and sad look on his face. He holds in his hand a dispatch just delivered from the radio shack.

"He's gone," he says quietly. "The Chief has gone. President Roosevelt is dead." The ship is strangely quiet. Somehow we have lost not just a president but a captain, our Chief. Most of the young crew have never known any other president than Roosevelt, now in his fourth term. They knew he would win the war. Now F.D.R. is dead, just as we are about to succeed.

I go down to the wardroom to get a cup of coffee, properly known on our ship as mud. Leroy "Bull" Kelley, a senior warrant bos'n in the U.S. Navy, is sitting there, alone. Kelley has been in the Navy for over twenty years; was wounded in the Japanese attack on Pearl Harbor; and had two ships shot out from under him in the Pacific. He is the toughest man and best bos'n I have ever known. Now tears stream down his face. And down mine, too.

Now I am a professor, who no longer wears a uniform but tries to teach a new generation that seems unable either to picture or fully understand the Roosevelt years, and what F.D.R. meant to this nation. If they have a tag for Roosevelt's New Deal, most of them would say it is Big Government. Look what that Big Government accomplished! Who would have us return to Hoover's philosophy?

Little wonder, then, that today's students find it hard to understand that the patrician Roosevelt was the savior of American capitalism, and the good life they take for granted. With his famous alphabet soup of new agencies, working closely with Congress, Roosevelt restored confidence, renewed vitality, and set the wheels humming again in the 1930s. The New Deal changed American life—and all of us today are among its benefactors.

Intellectuals, academics, and media joined Roosevelt and his daring programs in overwhelming numbers. The political transformation brought about by the New Deal was, in the words of historian Norman Cantor in his book *The American Century* (1997), "profound and long lasting." It moved the mass of American academics and intelligentsia to the left of center position where it remains.

Roosevelt's policies remain controversial today—as do those of the other great presidents who are his neighbors in the Tidal Basin. Their reputations fluctuate like the stock market. Just now, as Roosevelt's stock goes up, Jefferson's seems to be coming down. Emerson explained why in a memorable sentence: "To be great is to be misunderstood."

The tilt of F.D.R.'s head and the boom of his laugh sustain us still, as the Ship of State finishes the century in perilous waters. The memory and message of our pragmatic philosophers, inventors, and presidents is our great hidden resource. There are other things, too.

We have had great achievements in our popular arts, the central core of all popular culture since our ancestors lived in caves and buried their god-leaders in tombs. The electronic arts are largely the product of our century—radio, motion pictures, video, television, computers, high-speed printing. Other forms, far more ancient, still sustain us—architecture, music, and dance, for example. We turn to them to complete this miniature mosaic of modern times. We begin with rock 'n' roll, paying homage to the century's most influential entertainer, the incredible Elvis, and finish by joining in the fun, doing the Twist.

Elvis the Incredible

He is dead and gone—or is he? Elvis the King still seems to reign. He is a landmark figure in the modern world.

It's been over twenty years since Elvis Aron Presley died on August 17, 1977—more than enough time for most rising stars to wane and pop singers to pass into oblivion. But not Elvis. He is enjoying a life after death even more spectacular than his earthly one. In a field where most fame lasts fifteen minutes, his postlife has lasted many years and is gaining around the globe. Even those of us who study and write about him are baffled, and find the question hard to answer.

Two Ts spring to mind: timing and talent. He came, like all charismatic leaders, in the "fullness of time." He burst on the American scene in the 1950s when comfort, complacency, and conformity abounded. Elvis was the rebel, in looks, sounds, and actions. His hips swiveled—his fans screamed. He combined rhythm and blues with country music to produce a new style that was neither black nor white, but electrifying. He had a slight sneer and a pompadour hair style, and did he move! From eyes to thighs, from lips to hips—Elvis the Pelvis took America's youth by storm. He was a precursor of two of the monumental events of the 1960s—integration and the sexual revolution. Many parents didn't like what they saw in Elvis, but the kids—and television—won out.

Good luck and shrewd managing helped. A small-time entrepreneur, Sam Phillips, who owned Sun Records in Memphis, saw Elvis's unique talent. Colonel Tom Parker ("Colonel" was an assumed title) saw big bucks, and got Elvis a contract with RCA records. He hit a home run with the recording "Heartbreak Hotel" in 1956 and the winning streak never let up.

At first, TV networks refused to let the camera show this twisting tornado below the waist. But Ed Sullivan turned him loose and the frenzy followed. Paramount Studios gave him a seven-year Hollywood contract. Elvis went from tour guide to sheik to cowboy to ex-con in movies that got little critical acclaim but ever-growing fame; the world loved Elvis.

In 1968 Colonel Tom Parker redesigned him, dressing him in black leather and putting him on the stage. Even though he was older, his young fans went wild. Elvis became paranoid. Afraid he would be assassinated, he surrounded himself with bodyguards. He persuaded President Richard Nixon to allow him to carry a gun aboard airplanes—and owned thirty-eight guns when he died.

Elvis thought the Beatles were communists, and craved his favorite food—fried peanut butter and banana sandwiches and deep-fried pickles. He started to forget his lyrics. No matter. He got fat. They still loved him.

But Elvis couldn't stand the glare and glitz. The death of his mother also left permanent scars. He put on extra weight, overdosed on prescription drugs, and died of heart failure at age forty-two. In one of his last TV performances, he sang with great emotion: "I did it my way." He did. In so doing, he changed the face of popular music, and perhaps even of cultural history.

Does the story end there? No, in a sense it just begins. RCA Records officials announced in 1997 that since his death they had sold 400 million Elvis recordings outside the United States, with an increase of 61 percent in the previous twelve months. This brings the grand total to well over $1 billion. A new platinum boxed set is yours for $69.95. There is also a new toy linking Elvis with the Barbie doll—it will be called "Barbie Loves Elvis." Does that surprise anyone?

Meanwhile, the market for memorabilia and trinketry booms. Forty-year-old Elvis tapes have been found and released, as well as various "lost interviews." You can buy a replica of the shirt Elvis wore for Crown Records, another of his guitar, a Graceland tour book, a huge *Elvis Remembered* book which is "just screaming for a coffee table," and on and on. You don't have to leave home to get any of these. There's an 800 telephone number. Soon we'll have a new word for it—Elvismania.

Or perhaps you *will* leave home, joining over 800,000 tourists who visit Elvis's home every year, which places it second only to the White House. We have no accurate count of fan clubs around the world, but they are especially strong abroad in England, Germany, the Netherlands, South Africa, Japan, and Singapore.

In my own travels, I am astonished at the fervor of these clubs, and the number of Elvis look-alikes and impersonators. There are surely thousands of them. A fan in the Netherlands has found a statue of Elvis that he claims sheds real tears, and many of the faithful come to worship and adore. Obviously this imitates the miracles of the Virgin Mary and her sacred tears. Saint Elvis?

How has Elvis been able to captivate so many for so long? No one can give an exact or complete answer. But his story—a poor unknown young man from Mississippi fighting his way to the top, and succeeding, is the great American Dream. Other obvious examples are Abraham Lincoln, Walt Whitman, and Charles Lindbergh.

His unique gift was his music, allowing him to express himself in sounds as well as words. With this he transcended race, gender, and even nationality. Nobody put as much into a song (be it a good or a bad one) as did Elvis. As one of my students put it, "He sings with his heart." Another student, a staunch believer, has just brought me a picture of Elvis on Mars. Who said the sky is the limit?

There is not only emotion and sincerity, but also vulnerability in his music. You feel the passion, searching, and inner turmoil whether or not you like the music. Because it is readily available long after his death, this music universalizes his talent.[1]

And his career personifies the American Dream. Born in abject poverty, close to the soil and his roots, he demonstrated the potentiality of our nation and culture. His power grows as we enter a new century. Long live the King!

105,529
DESIGN FOR A BOTTLE
Eugene Kelly, Toronto, Ontario, Canada, assignor to The Coca-Cola Company, Wilmington, Del., a corporation of Delaware
Application March 24, 1937, Serial No. 68,391
Term of patent 14 years

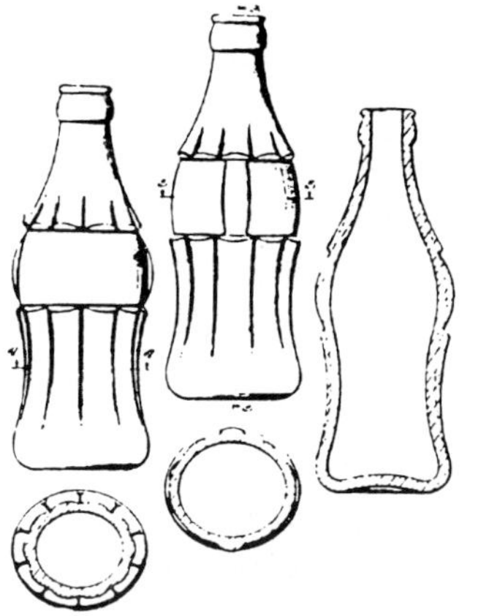

Evolution of Bell Telephone Emblem Over Years

FIGURE 1. A glass bottle (Coke) and a circled bell (Bell System) have become world icons of popular culture. Everyone knows that Coke is "the real thing"—and that with the telephone, you can "reach out and touch someone." Courtesy of Craig Gilborn, Winterthur Museum.

FIGURE 2. The Brick Woman is my symbol for Bangladesh. With a heavy hammer, she smashes bricks into small pieces, day after day, endlessly.

FIGURE 3. The distant past and the present meet on a Bangladesh rickshaw painting. A young Bangladeshi from the real streets poses above the imaginary streets painted on a rickshaw by a folk artist.

FIGURE 4. East meets West in Beijing's McDonald's. Ronald McDonald is very happy. Photo courtesy of Brad Trask, McDonald's Corporation.

FIGURE 5. Surely we are in Arcadian Greece? No—on the lawn of White Sulphur Springs, West Virginia. True enough, the temple is pure Greek, and the goddess on top is Hygeia, who brings good health. The invitation is mythic.

FIGURE 6. Policemen throughout the world—in this case, India—direct traffic under advertisements for American products—in this case, 7UP.

FIGURE 7. The folk art statue of George Washington holding an American flag sits atop Washington and Lee University in Lexington, Virginia. The pinnacle has the "wall of Troy motif" favored by early Greek artists. Folk, elite, and popular culture merge, as they do more and more in today's world. Photo courtesy of Frank Parsons, Washington and Lee University.

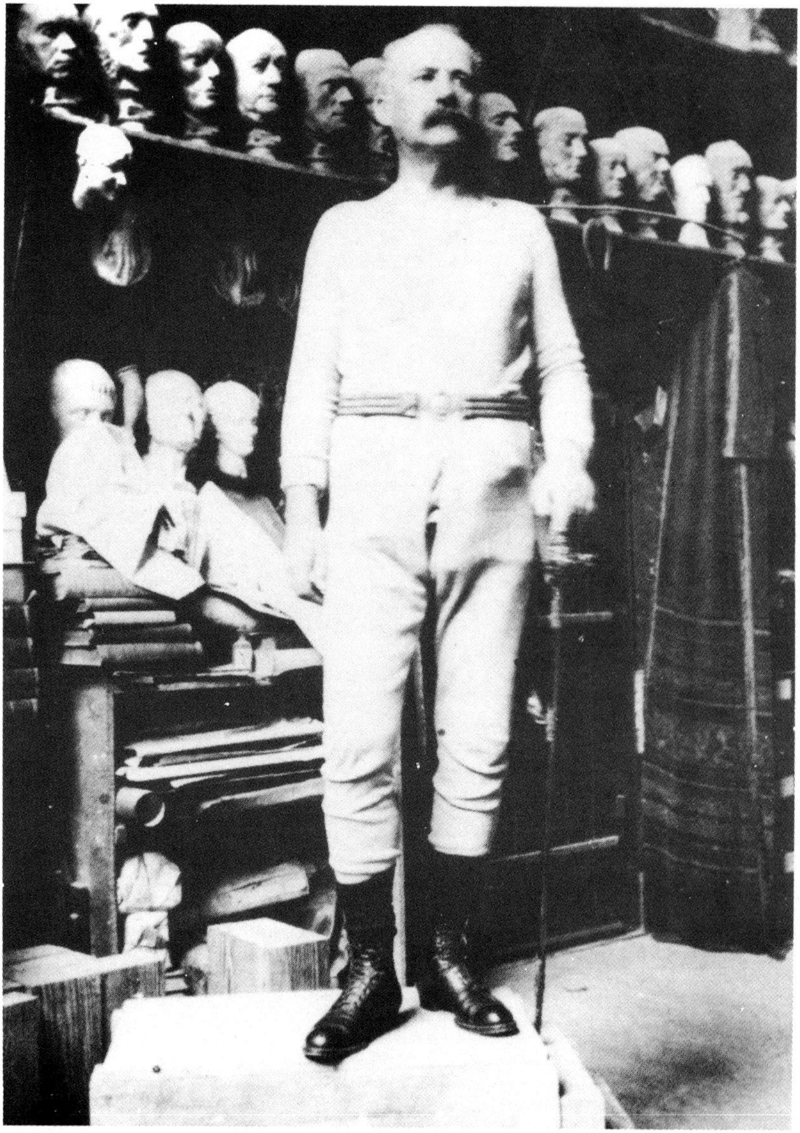

FIGURE 8. In the Richmond studio of sculptor Edward Virginius Valentine, Robert E. Lee, Jr. poses for a sculpture of his famous father, General Lee. General Lee's hold on the Southern imagination has been a phenomenon of popular culture. Photo courtesy of the Confederate Museum, Richmond, Virginia.

FIGURE 9. Why do we go to war, kill our neighbors, and wreck nations? Because it is our DUTY. We have known this for thousands of years. This poster, published by London's Parliamentary Recruiting Committee in 1916, helped keep bloody World War I going. Rather than saving Europe's global supremacy, the war destroyed it. Courtesy of David Allen & Sons, London.

FIGURE 10. The Golden Arches have found a place in 120 nations throughout the world. They celebrate a food named after a German city (Hamburg), by a company named after two Canadians, purchased and globalized by an Eastern European immigrant to America. Is theirs the greatest story of twentieth-century popular culture? Has the church steeple been surpassed by the Golden Arches?

FIGURE 11. All cultures must come to terms with God, tribe, and family. In an age of computers, satellites, and the Internet, there are hundreds of small towns—such as this one in Virginia's Shenandoah Valley—where those items are visible: the church steeple, the American flag, and the woman sitting on her front porch in a rocking chair, sewing.

FIGURE 12. For countless centuries, the camel, ox, and horse carried civilization forward on their backs. The memories last even when new forms of transportation take over.

FIGURE 13. S. W. Jackson kept this general store in Albermarle County, Virginia for fifty years. He carried most of the things people needed in the first half of the twentieth century. Now, his grandchildren and great-grandchildren go to massive shopping malls, where thousands of items await. However, at the malls, there are no S. W. Jacksons to talk to and tell local stories. Popular culture has become mass culture.

FIGURE 14. James Montgomery Flagg drew this stern Uncle Sam, seeking recruits for WWI. "I want you!" the caption read. Few pay any attention to his stern command now.

FIGURE 15. Time has been the essence of popular culture, from cavespace to cyberspace. As the twentieth century ends, we have trouble deciding what we think of time—or just how it rules our lives.

Twisting

Whenever people anywhere rejoice, they dance. No matter what the time and place, they feel the need to use their bodies to express their emotions. Victories, births, weddings, homecomings, centennials call forth dancers. Sometimes dances seem to give an age its signature—the minuet in the eighteenth century, the waltz in the nineteenth century. But does any one dance really capture the mood of the twentieth?

"When I play on my fiddle in Dooney," William Butler Yeats, an early twentieth-century poet, wrote, "Folk dance like a wave on the sea." Henry Dobson also used the fiddle metaphor in his poem "Cupid's Alley":

> O, Love's but a dance
> Where time plays the fiddle.

I want to choose a popular American dance for one of my vignettes describing modern times—though the folk fiddle doesn't work for American popular culture. There are many candidates. Square dances and clogging were favorites in the country. Break dancing was an African-American specialty. Both the dance and the name "jitterbug" caught the spirit of the 1940s. I ended up recalling the Youthquake of the 1960s and choosing the twist. At a time when we "let it all hang out," it was our Brave New Whirl.

Oh how we danced, pranced, and shimmied! The ultimate spot was New York's Peppermint Lounge, and the king of the scene was Chubby Checker. You can't get much more pop American than that.

Our parents, matrons, and mentors were horrified. Not even the Russians, who drove us under desks and tables in practice atomic bomb drills, came in for such condemnation. Many parents forbade the twist individually, and some communities collectively. Such goings-on were, we were told, erotic, barbaric, degrading, satanic.

The same epithets, my mother admitted, were used when she and Dad danced the Charleston. So of course we had to twist. At a time when there were no wars to challenge us, the twist was our great adventure.

Even at this late date I shall not debate its merits or demerits. Who can say? Whether the demure American girl, in her silk stockings, girdle, and necklace sacrificed genteel love and romance for sex on the move, I am not prepared to say. But my father admitted, having drunk his before-going-to-bed Scotch, that moving sex was pretty popular in his day—and Granddad's. "There's always been he-in and she-in around here," he confided, "only other people are doing it now."

Doing it—cakewalk, camel walk, eagle rock, shimmy, black bottom, Charleston, snake hips, lindy hop, shim sham, truckin', Susie Q, Shorty George, peckin', boogie woogie, Big Apple, shag, sand, applejack, mambo, cha cha cha—just where do these new dances come from? And what of tap dancing, which seems to have a particular and enduring home in America? Why are the others suddenly popular, only soon to become part of the passing parade?

The tempo of the times is mirrored in its arts and its literature. We have been thrust into urbanization, mechanization, specialization, synchronization, then into cyberspace. Theories of energy, matter, and time are in flux. The certainties of Newton died in 1905 when Einstein published his theory of relativity. Suddenly most of our notions of the universe had to be revised or junked. So did our old ideas about music and dancing. Reality took on a variety of new profiles, and they wedged themselves into our minds and hearts.

Hucksters and pretenders cropped up everywhere. What cultural revolution has ever swept a people without confusion, pretension, and vulgarity? Who can draw the line between experimental and eccentric, new and neurotic, significant and silly? Certainly no one would be wise to defend all that is new—or to condemn it all, for that matter. Let time sort it all out.

As our various arts show, the pioneers of our times are Prometheans, putting aside the old slogan of Onward and Upward, insisting that we are moving instead from Progress to Perplexity. All efforts to explain why this describes contemporary America fall short. We are left with the memorable metaphor of F. Scott Fitzgerald. France has a land, he said, England has a people, but America is harder to utter. It is a willingness of the heart.

And, if one examines its dances, a willingness of the arms, legs, and feet. Strike up the band, kick up your heels! Dance on, O Ship of State!

* * *

ENDNOTE

Much modern information and knowledge comes not from books but from newspapers, magazines, films, television, and the Internet. At the same time, there are books that help prepare us for the sound and fury in which modern people live.

Were I to choose one basic book of this sort, it would be R. R. Palmer's *A History of the Modern World*, used as a text for years in many colleges and universities. My favorite edition was published in New York by Knopf in 1996.

An excellent European counterpart is Henri Lefebvre's *Introduction to Modernity* (London, Verso, 1995). He struggles with a basic question—"What is modernity?" So does Britain's Hugh Thomas, in *World History: The Story of Mankind from Prehistory to the Present* (New York: HarperCollins, 1996).

Ours is the age of many things—science, technology, revolution, and triple packages (such as gender, race, and ethnicity). All these areas have produced articles and books galore. My own inclination would be to call ours *The Automobile Age*, as chronicled by James J. Flink (Cambridge, MA, MIT Press, 1990).

Music

Music is a universal language, and it has always flourished in popular culture. It has become the hallmark of the young, and if there is a universal language, it is pop/rock music. The music of the young, changing with lightning speed, has become not only the mirror but the triumph of our children—and the despair of many of their parents.

The one form that has been most influential throughout the world is jazz. It had humble beginnings in New Orleans brothels—some think the word "jazz" is derived from the word "ass." If the derivation is obscure, the impact is obvious: jazz turned its back on the world of decorum, control, and restraint. By superimposing syncopation, polyrhythm, and riffs, jazz offers unending variety, dissonance, and a tension of counterpurposes. Traditional European rhythms are chordal or vertical, while jazz is linear or horizontal. Its appeal to many Europeans was enormous. It still is today.

Some politicians began to ask: Why has jazz won more young converts than either the Christian Church or the Communist Party in the past fifty years? Might it have been its ability to reflect our times and inner emotions in ways that traditional music can't? Did it open doors that had long been closed? Some theologians believed it might. Wilson Wade, for example, said we may find in jazz the long-awaited symbol for the renewed affirmation we so desperately needed. A Roman Catholic professor, Father James Kennard, observed that to swing is to affirm.

Liking jazz is one thing; defining it is another. Here are some suggestions: Jazz is a lazy summer afternoon. Sand between your toes. Stinging nettles. Not letting the right hand know what the left hand is doing. Bourbon on the rocks. Sleeping on foam rubber. Kicking your bare toes on a piece of flint.

Jazz is an essential ingredient of popular culture: a recoil from the terrifying changes, a sound to help heal the alienation that separates and corrodes.

* * *

But can this be said of the new age of music, which came in with the Youthquake of the 1960s and has cut like a buzzsaw into our music and our psyches? Has the Age of Jazz been supplanted by the Age of Rock, Roll, and Rap? Does the new electronic music continue a great tradition, or destroy it?

Harmony has given way to dissonance, delight to defiance, mild to wild. As the decibels went up, the melody and lyrics tended to disappear. In came tribal shouts and erotic grunts. There is a parallel here with modern art. Many artists have given up painting still life, landscape, and grandmother, and turn for themes to anger, angst, and pornography. Some even use graffiti from toilets as themes. Why not? they say. Why? some of us answer.

Not only the instruments, but also the voices accompanying them have changed. Listen to the top performers, and you will hear that synthesized nasal whine, that rasping yell, moving swiftly from Michael Jackson to Madonna and on through the years. By the time you read this, all this will have changed. New sounds and new faces will be on the charts.

In some ways it is futile for us "Golden Oldies," as students call my generation, to write of these things. We are on the outside looking in. Punk rock and heavy metal featured the body-music ethos, and tended to make pop music into a weapon of social assault. Racial and ethnic slurs abounded. The feedback is so loud it can literally destroy one's hearing—a kind of aesthetic toxic waste.

The focus of the multibillion-dollar music industry has switched from composition to engineering. The new technology opened up new ways to store, edit, and recombine sound. All this diminished the value of live musical performances, which tended to emulate the mechanized sonic perfection of records. Many now try to reproduce the look and style of video clips, including lip-synching music already canned. So who is singing what? Who knows?

One need no longer go to Nashville, Motown (Detroit), or New York to record. Those who can afford it build a home studio, armed with a computer-laden arsenal of sequencers, synthesizers, and sound effects that make songwriting and sound engineering almost synonymous.

This does much more than change the actual sound of the music, and the technique involved. It tends to change the whole process of music making from a communal working together to one person working alone. Even pop records using a musical ensemble are not what they seem. They may well be put together by compiling different instruments, track by track, in separate recording sessions. The drummer may never see the trumpet player—and neither will see the singer.

The parallel between this and our lives at the computer screen is obvious. Virtual reality has reached out from the screen and taken over the very lucrative music business. Success is no longer measured so much by talent as by electronic gadgetry.

What might Bach or Mozart say to all this? What does it say about our popular culture?

Postmodern

Postmodernism: something old, something new, something borrowed, something blue. But just what is it? Where did it come from?

There are as many definitions as there are advocates. A common thought is that we must move beyond modernity. Why? Because of the desperate need to accept the presence of volatile and unstable elements. We must put Descartes and his cold rationalism behind us, and start feeling again. In this sense, it is a rebirth of nineteenth-century romanticism mixed with a dash of anarchy and antiestablishment. Enter, on the pop music scene, the hip-hop of disc jockeys, who change styles as fast as tunes on turntables. Live the zap-around-the-dial mentality. Add porn and corn to the Internet. All the world is at your fingertips, so use them. Break all the old boundaries. Zap-rock!

The 1990s found the modern world in a style free fall—borrowing, bending, crunching, and bunching odd parts to make the Old Boys scream. Out with the old, in the the new.

I got an earful of this when a new rock group called Southern Culture on the Skids invaded our campus for an "end of exams blowout" concert. The name was well chosen. Loud and lusty, they blasted us with music they described as "hillbilly-surf-mexicali-rockabilly." One of the spectators, having enjoyed a good amount of artificial stimulant, called it "low-down-dirty thunder that's straight-outta-the-sewer blues."

Reactions to the concert were mixed. Everyone conceded that this was indeed postmodern, and post- lots of other things. We were not surprised to discover later that the lead singer had been inspired by watching bugs.

What is behind this radical break in the way some of our young people think, sing, perform? Instead of condemning, we might try understanding. Are we witnessing the breakdown of what used to be called "reality" in the face of revolutionary new technology?

Can we see the same tendency in other contemporary forms—such as the new science-fiction writing?

I think we can. Stanislaw Lem is a pioneer in the field. He believes personal identity and objective reality are gone forever. In *Futurological Congress* (1996), Lem's hero is caught in a cloud of hallucinogenic tear gas. He perceives future societies organized entirely around drugs. "Pharmacocracy" has replaced democracy, so that the whole world has become a virtual reality. The distinction between manipulated and natural feelings has simply ceased to exist. We have decided to stay forever in Plato's cave.

Another interesting example of post-modern speculation is the work of Philip Dick. In his book *UBIK* (1991), the ancient boundaries between egos collapse, and biologists discover what is called "half life." The dead are not really dead. They are frozen in "cold pack" (as Walt Disney is reputed to be) and are electronically connected with each other—and with the living. In effect, what we now call death has been eliminated. Paradise regained!

The main characters in *UBIK* are dead throughout much of the story, yet "tuned in" to the living. Sometimes they start to fade out. Science has a solution—a special spray can bring them back to half life. If it doesn't work could you linger in quarter life? A new version of purgatory?

Absurd science fiction, one might say. But might all these tales be metaphors for the hazards we face in the days ahead?

Except in small pockets of earnest students of English or philosophy, postmodernism hasn't made major inroads at universities. In typical fashion the students, politically indifferent and fed up with false rhetoric, have shortened postmodern to "po-mo," and put it aside.

Not so with the faculty and intellectuals, who have invented a whole new lexicon of words and phrases. They are the egalitarian avant-garde, shooting arrows from behind their academic walls. They speak of the rave scene, moral panics, and implosion, then move on to greater polysyllabic jargon such as intertextuality, appropriation, neopolysemous, macropolitical, hegemonic, valorization, deconstruction. They sometimes fall back on terms that emerged in the 1960s and have taken on a new life now—Eurocentric aesthete, or cultural imperialism, which posits a process of imperialist controls, and blames the usual suspects—corporations, power politicians, Anglo-Saxon males, etc. This the students (especially some feminists) can find more to their liking.

Nonetheless, the cult figure for many postmodernists, Friedrich Nietzsche (1844-1900), is a DWEM (Dead White European Male), and Jean François Lyotard, who wrote *The Postmodern Condition* (1984), the bible for the faithful, is at the moment at least alive. While Lyotard is alive, is modern history itself, as some postmodernists claim, now dead? Lutz Niethammer raises the question in *Posthistoire: Has History Come to an End?* (1992). The implication is that things are so new that all the old concerns, questions, and assumptions are outmoded. We have entered the dark regime of conceptual history. Now all the historian can do is to gather together some little stones, without knowing if they form part of a mosaic.

I side with Henri Lefebvre, whose *Introduction to Modernity* (1995) takes quite a different approach. With or without dignity, in plush or in tatters, however rapid and noisy, what we call modern marches on. The modern world is self-evident—in painting, music, science, and political life. Modernity ought to be obvious to all but the myopic, dogmatic, or completely blind.

In the 1980s postmodernism was at the forefront of radical chic, but within a decade, it began to lose momentum in key areas, such as the arts. Having attracted wide attention by announcing the death of the ideals and principles of modern art, it began to repeat the mistakes of the earlier art world and show the same signs of fatigue. Dealers and critics complain of a lack of direction and purpose in the new art. They find a lack of spiritual values and depth. Like the rage television series of the 1990s, *Seinfeld*, it seems to be a show about nothing.

Drawing heavily on French poststructuralists such as Derrida and Lacan, postmodernists attacked stereotypes, repetition, and media domination—but have ended up stereotypes themselves. They have become like the very things they set out to oppose. Hence this headline in the September 16, 1990 issue of *The New York Times*:

As It Must to All, Death Comes to Postmodernism

* * *

What's in a name? Everything. Without a name we are nothing. How can a person or thing exist without a name (and for anyone in today's world, a social security number)? Everyone plays the name game—not just for themselves, but for everything they see, hear, taste, touch, and buy. No name, no nothing. Desperation. Let me give you an example. What's the right name for the generation of the 1990s? The twentysomethings? The children of the Baby Boomers who have taken over Washington? Generation X?

Don't be surprised if you can't come up with a quick answer. My research and questioning indicates that there isn't one. There is no generational logo.

We haven't had this problem for decades. If you were born in the 1930s you were a Depression Baby. Everyone in the 1940s knew about G. I. Joe and Rosie the Riveter. We called the well-heeled and complacent children of the 1950s the Silent Generation, and their daddy, the Man in the Gray Flannel Suit. Then came Ginsberg's *Howl, The Greening of America*, and the Youthquake, teeming with beatniks and hippies. Things quieted down after the orgy at Woodstock, and the Weathermen and Black Panthers came in from the cold. We talked of yippies in the 1970s and yuppies in the 1980s. Tom Wolfe coined a phrase for the 1970s that stuck—the "Me Generation." They loved their BMWs, brie cheese, white wine, and eventually Ronald Reagan. Working in and out of the government, they gave us the greatest national debt in history.

But the 1990s? There's the problem. Not that we lack for suggestions. Call them the Baby Busters (as in "from boom to bust") or the Baby Boomerangs (coming back home without a job). What about the Lost Generation? No, Hemingway slapped that on the 1920s. The New Lost Generation? We can't even remember the old one.

A little probing turns up lots of other suggestions—the Cable Generation, the Brady Bunch (for the rerun they love to watch), cyborgs, videos, latchkey kids, mall rats, downbeats, burnouts . . . the list goes on and on. People who make big bucks from coining the right names are frantic. Anyone who sells or advertises anything gets into the act. They point out that over 80 million Americans were born between 1961 and 1981. What's in a name? Everything. How do we create some community and identity, since the age-group ID is a cultural imperative?

Perhaps we should ask the twentysomethings themselves. Here are two answers. Rachel, twenty-four, says, "We are sort of caught in the big muddle of trying to see every side of everything and not being able to move forward." Multiculturalism? Or this from a twenty-eight-year-old law student:

> We are caught behind a defining generation—we must settle for junk culture as a unifying symbol. We couldn't rebel against Vietnam. We missed the eighties boom and all the good jobs were taken. So we invented this slacker psyche and in my case went to law school.

Then came a suggestion from the nation's number one Baby Boomer, Bill Clinton. Hard at work on his national sales blitz, a public relations campaign to pressure Congress to support his economic agenda, he came up with a politically correct suggestion. Why not move from the "Me Generation" to the "We Generation"? Isn't it time "we" came together, to share the burden and the pain of renewal?

It's too early to say—but the fast-track lobbyists and special interest groups are nervous. Does Clinton expect the successful Baby Boomers to cut back on wine and cheese? Or Congress to give up that ultimate pork barrel, the soft money? Or the Golden Oldies to cut back on their social security checks? Or any of us to pay a whopping tax at the gas pump?

Maybe if we wise up WE will accept the WE.

To get to "we," we have to curb "E"—the obsession with entertainment, which has spread like the Exxon oil spill off the Alaska coast, blanketing our whole culture—religion, journalism, public policy, education. In a recent poll, the number one reason college students gave for cutting classes was that the class was boring. Did anyone ever say that tackling a tough mathematical problem or translating a foreign language was supposed to be fun and games? Do our students spend thousands of dollars to get professors or song-and-dance men? And how "entertaining" is the real world in which they must struggle to find jobs?

Naming the new generation, and being sure they are educated and not merely entertained, is only a small part of our problem. What role shall this New America play in the Age of Globalism?

The size, spectacle, and profits of this new world enthrall and seduce us. Millionaires become billionaires, mergers multiply, and the stock market goes through the roof. We go faster and faster, work harder and harder, and keep losing the car keys.

No matter where we go or what we spend, we have to live in two soils. The first is physical, the second is spiritual. To endure, a civilization must plant seeds in both soils. It must draw not only from the land, but also from what love and deep feeling makes of those who cherish and live on it. Here is the sinew of the soul, without which big bucks and big deals mean little. We must define not only our new generation but our new goals and priorities.

As this century and millennium ends, there is a general (if sometimes grudging) agreement that America is a superpower with a super style that is sweeping the globe. Like Britannia before us, America now rules the waves, the markets, and the menus.

If we count dollars, aircraft carriers, nuclear submarines, bombs, and bombers, that claim cannot be contested. The Soviet Union has collapsed, and other possible contenders are only beginning to emerge. We have the power. With power go the perks: predominance in media, movies, music, fast food.

Do these and other American items mean that we also have a world-class super style? Is this our real secret weapon as we enter the twenty-first century?

Those whose power and perks have faded think it might well be so. Britain's Lord Hugh Thomas, in his recent book *World History* (1996), concludes that this American domination offers the prospect of a better world than any other that can reasonably be imagined. This takes into account that the world might copy our style of government, entertainment, fashion, even food. He might be right. The same month his book came out, McDonald's opened a fast-food restaurant in its one hundredth overseas nation.

Not everyone shares Lord Thomas's optimism. In another study, *Europe: A History* (1995), Norman Davies warns that a mindless materialism of a transatlantic homogenized culture could have insidious, even disastrous, consequences. People, as well as products, might become homogenized. Many share both views, and even more come down in between.

We must examine both the problems and prospects of an American super style in the global village. It is a complex and multifaceted puzzle, not amenable to short-sighted solutions. If we can raise some questions, and encourage others to do the same, we will at least have moved the argument forward.

In the final analysis we suspect Woodrow Wilson, that great internationalist, was right when he said you can love a country or a world if you begin by loving a community.

Two soils, two strengths. One strength is of the earth, the other of the human spirit. May they merge and maintain us in the tumultuous days ahead.

Some Postmodern Questions

In the fast-changing world of cyberspace, names come and go. What was Cyburbia became Cyberville in 1998 in Stacy Horn's trendy book *Cyberville: Clicks, Culture, and the Creation of an Online Town*. Her town is Manhattan, a small place where thousands of members log on daily to hang out and talk, gossip, and gripe . . . where unexpected bonds are forged in all but pure air. She then prints page after page of Internet "no-holds-barred cyber-symposiums on . . . romance and (of course) sex!"

You can get the latest buzz from people hiding behind nicknames such as Maggot Boy, Bite the Wax Godhead, Cafephreak, Lana Tuna, or A Little Black Cloud in a Dress. Or you can hear some very startling questions from Groupie Girl, such as "Do you get together and touch each other's monkeys?" Or (from another source) "Why is it that a blow job doesn't seem to count unless one swallows the hallowed fluid?"

All of which sounds like the talk of teenagers in the school locker rooms. Who buys and praises such stuff as this? Abraham Lincoln knew. His "Judgement on a Book" is recorded by G. W. E. Russell in *Collections and Recollections:* "People who like this sort of thing will find this is the sort of thing they like." Nothing wrong with that. What irritates is the self-praise and hype of Ms. Horn, who claims "This shows the truth of our lives." What truth! What towns!

Cyberspace, like postmodernism, has become a growth industry. In the nineteenth century, critics said that everything that came loose went West. In the late twentieth century, everything that comes loose gets into cyberspace—and a dozen hucksters wait to extol and expand it. Sometimes even multimillion-dollar satellites stop working. We cite the case of Galaxy IV.

Of course there are enormous and obvious advantages that come with the new technology, and they must be given their due. It's hard to see how we would predict tomorrow's temperature, get a few rocks from the moon, reserve an airline ticket, or find out if there's water on Mars, without the new gadgetry. But considering the billions it takes to accomplish these feats, one has to ask: "Is it really worth it?" How did we get along for thousands of years *without* those rocks, and the answer about Mars' water supply? When does the information overload become intolerable?

These postmodern questions must await postmodern answers.

We are standing on a new virtual Dover Beach, reminding us of Matthew Arnold's real one:

> And we are here as on a darkling plain
> Swept with confused alarms of struggle and flight,
> Where ignorant armies clash by night.
>
> "Dover Beach"

We must not come as sentimentalists, obstructionists, or Luddites. Just as King Alfred standing on the seashore could not halt the incoming waves, we know that our culture is committed, body and soul, to what many believe to be a New Utopia. The changes wrought in recent years are dramatic, drastic, and (in some cases) irreversible. The consequences of these changes have only begun to emerge. No one can predict what they will finally be.

We don't know where we're going—but we're on our way—like the old Red Ball Express, hurtling down the track with the throttle wide open. Theories of energy, matter, space, and time are in flux. What was once simple has become incredibly complex. Everything nailed down has come loose. Information fights disinformation, smut doubles with glut. Millions of terminals send messages with the speed of light. Anything and everything goes. Data data everywhere and not a chance to think.

Communication is becoming "compunication"—the linking of computers, fiber optics, satellites, and the new Toy of the Month. Once all the world was a stage. Soon it will be a mall.

Who benefits from new technologies, mergers, buy-outs? Will a new overclass usurp much of the power? Will those unfamiliar or hostile to the new electronic wizardry become a new underclass—underemployed, undervalued, and unemployable?

Revolutionary changes are always frightening, especially for those raised in the earlier print culture. We know how earlier technologies—such as the automobile, television, nuclear power—brought grave concerns. We also know that those concerns were justified. The automobile, television, and nuclear power plants have raised problems few if any anticipated or understood. The earlier naive euphoria has disappeared. We are perplexed and polluted, bewitched and bewildered.

We can't halt the flow of information, nor deny the many obvious postmodern advantages. New wonder drugs work; we have advantages undreamed-of even a generation ago. We also have new questions.

Is big better? Is biggest best? Who will arbitrate and regulate the Electronic Revolution and the media? Who will protect our basic privacy? What is too slanderous, libelous, or pornographic to be available to everyone? Do our doctrines of free speech and free press fit today's world?

Where to draw the line between the silly and the significant, the truth and the hype, the new and the neurotic? Add these to our postmodern questions. Don't expect easy answers.

* * *

Death has not and will not come to the underlying causes that birthed and shaped postmodernism. It was born out of a desire to change—to discredit—to react against the present. This urge and ferment for change—a kind of metamorphosis—goes on constantly. Key words are "new," "radical," and "post." There will always be a New Deal, a New Freedom, a New Frontier. They are part of our heritage, and of our dream.

So well may we ask, with Shakespeare, what's in a name? Doesn't a rose by any name smell just as sweet? A baffling question. What causes "isms" to become "wasms"? Where do the new names, the new isms, come from? Who named today's twentysomethings Generation X? Will the tag stick, or is it already changing?

As mosaic makers we shall explore that question, along with changing concepts of time, and basic issues now arising about postmodernism. Piecing these together might cast some light on the murky waters in which we now swim.

ENDNOTE

If anything, "postmodern" is harder to track and pin down than "modern." Here too we face many electronic sources and an information glut. The prevailing cliché is that we have "massified." If that be true, a good introduction is the popular anthology of Stan L. Wilson and James R. Jameson: *Mass Media—Mass Culture: An Introduction* (New York, McGraw-Hill, 1997). Another favorite topic is the millennium. Ray Browne and I set forth our thoughts on that in *Preview 2001+: Popular Culture Studies in the Future* (Bowling Green, OH, Popular Press, 1996).

If as I contend the best insights on postmodernism came years ago from Marshall McLuhan, his best book dealing with that is *Take Today: The Executive As Dropout* (Don Mills, Ontario, Longman Canada Limited, 1972). From there the trail leads every which way.

The only credible biography of McLuhan is Philip Marchand's *Marshall McLuhan: The Medium and the Messenger* (Cambridge, MA, MIT Press, 1998). The book offers clues to a genius who was all too easy to misunderstand.

Cyberspace

Cyberlore came down on us like a bolt from the blue. There were no computers in the world when World War II began. During that war we developed one for military purposes—a monster weighing fifty tons—to speed up ballistic calculations. Pandora's box was open. It may never be closed again.

"Cyber" is the Greek word for helmsman or guide. The word has expanded like the computers that are its trademark—into our offices, factories, homes, economy, and mythology. To many, the computer is an icon. To others, it is a threat. To everyone, it is a consuming new element of poplore.

Like many others, I struggle to place this "lore" into perspective, and to measure its impact. We are all too close to it, too involved with it, to make definitive and permanent judgments. My general attitude is skeptical—but that will have little to do with how we fail or flourish in cyberspace. As always, history gives us the one wise answer: wait and see. We got a brief glimpse of one potential danger in cyberspace in 1998.

* * *

In cyberspace, things change in a moment, in the wink of an eye. We found this out at 6 p.m. on May 19, 1998, when a single satellite, spinning uselessly 22,500 miles above Kansas, silenced 90 percent of the nation's 45 million pagers. Credit cards wouldn't work, radio broadcasters went silent, and chaos set in for many doctors, emergency vehicles, hospitals, weather meteorologists, law officers—even drug dealers—who were tied in to Galaxy IV. Thousands of cable TV systems and television stations had nothing to broadcast.

Pan-Am Sat, owner of Galaxy IV, assured us that we have "the best communication system in the world"—but had no idea why the super system failed. Did they see the irony of the two statements? Companies, traders, and ordinary people suddenly found themselves in the new Dark Ages.

No one knew for sure how the satellite could be fixed. The only recourse was to send out thousands of technicians nationwide who would physically reposition satellite dishes to a different part of the sky where a backup satellite was orbiting. The adjustment had to be precise—within an eighth of an inch.

The outage, by far the worst in the thirty-seven years of satellite service, suggested an analogy with the sinking of the *Titanic*. When built, the *Titanic* was state-of-the-art, top-of-the-line technology. It was "unsinkable." But sink it did, killing 1,500, when it hit the tip of an iceberg. No one could have guessed how much of the iceberg couldn't be seen.

Some wondered if the Galaxy IV disaster—deemed absolutely safe by what was considered the best technology in the world—forced us to relearn the *Titanic*'s lesson. Technology, so often our willing servant, can be a cruel master.

This time, we lost mainly pagers and broadcasts. That was serious enough. Pam Silvestri, working in Dallas's Transplant Hospital, wondered what would happen if, with pagers out, they missed a call about an organ transplant, in which immediate action was required. "I was freaking out," she said. "This is life and death."

One thing that died overnight was any hope for retrieving Galaxy IV. We have no indication that we will be able to use it. Good-bye to a $250 million sky-toy.

In older days we would have said those millions of dollars went down the drain. Now we might say they went out into nowhere. Surely this is a cautionary tale.

What similar failures, with worse consequences, might await us? Suppose, for example, the satellites controlling our defense missiles, unmanned planes, and Navy ship direction finders faltered? At a time when those of an enemy were working well?

This in turn raises a more fundamental question: are we the masters of our new technology—or will it master us?

* * *

Consider for a moment the Sorcerer's Apprentice. He used the Master's magic to get broomsticks to do his job, carrying water. It worked—but he couldn't stop the brooms. A flood resulted and disaster threatened. The angry Master returned just in time to save the day.

Mickey Mouse was the Sorcerer's Apprentice in *Fantasia*. I remember it vividly.

We don't have sorcerer's apprentices these days—at least not by that name. But try this possible substitute for the 1990s: technopoly.

This new popular buzzword (the title of a recent book by Neil Postman) restates the ancient axiom of Socrates—that the unexamined life is not worth living. *Technopoly* suggests that the unexamined technology is not worth having.

This fear of rampant and unchecked technology goes all the way back to the Luddites, seventeenth-century English followers of Ned Ludd who smashed spinning jennys to protect the handicrafts and cottage industries. In our own century it surfaced in books such as Aldous Huxley's *Brave New World* (1932) and George Orwell's *1984* (1949). More recently writers such as Lewis Mumford, Christopher Lasch, and Neil Postman have sent out warning signals. In *The True and Only Heaven: Progress and Its Critics* (1991), Lasch implies that there may have been some truth in John Randolph's definition of progress as a "luscious lie."

No one can deny that technology has done incredible things for us, in this era of silicon chips, space walks, AWACS, medical miracles, and virtual reality. The whole world was wired for the latest on O. J. Simpson. The question is, what has all this done *to* us? How much of the highly touted Information Revolution is built on ad hype, media distortion, TV ratings, and patriotic chauvinism?

Bill Gibson spent two years watching and analyzing 2,000 hours of TV tape—a single day's offering on a cable system with 120 channels. We all knew what he would find: the same thing, the same reruns, the same plots over and over and over. Television, he concluded, is indeed a drug—not a hallucinogen, offering new perceptions, but one guaranteed to repeat itself day after day. Only a rare program breaks through television's claustrophobic and centrifugal force. Now we are told we might end up with 500 channels!

But *Technopoly* has leveled its biggest guns at the computer. It does many things (processing, sorting, and storing information) extremely well. There are other things (thinking, feeling, creating ideas) which it can't do at all. Humankind's noblest ideas (justice, equality, compassion) are not computable. When disaster strikes, don't expect your computer to cry.

Yet many people—including those in positions of knowledge and power—do insist on humanizing and empowering the computer to lead the way. To them, computers have brains possessing intelligence. They even personalize the machine to the point of buying clothing for it. A computer in our building wears a business suit for meetings. MIT's John Pfeiffer has written a book calling the computer *The Thinking Machine*, and John Kemeny has forecast a "symbiotic evolution" of the human and computer species. We are hearing claims that computers are not only smarter but more durable than us; that they might become the dominant species of "life" on planet Earth.

Think of the implications. Homo sapiens, which nature evolved in two million years and Moses, Jesus, Buddha, and Mohammed gave their lives to redeem, will be doomed to oblivion by IBM and Macintosh.

Part of the problem is linguistic. Computer people are developing a language of their own, unlike any of the human languages. I always thought German was a difficult language (Mark Twain called a German passage "chloroform on paper") until I tried to read a computer manual. Shades of Orwell's doublethink! Once you learn one system, the model or indeed the whole system changes. We live in the electronic age of enforced obsolescence. Whatever you have today won't be good enough tomorrow. You can count on it.

Meanwhile, pray that your computer doesn't break down, reject disks, erase months of work, or get a virus. If such a calamity occurs—and we hear of them daily—don't try to fix it. Off to the professional repair service. In due time (and at considerable expense), it will come back with a message attached. One is sitting near my office now. "Internally Terminated. No Solution. At Ease. 1.0" Make of that what you will.

The Return of the Luddites

In the early years of the nineteenth century a self-styled "General," Ned Ludd, organized a revolt in Britain of traditional workmen against the new mill technology. The new industrial order had political power, but it didn't have the support of the workers. They believed that moral authority and community sanction were on their side, and that the new era reduced men to poverty, and women and children to a state of slave labor. Their cause was quixotic, and they were soon suppressed. But the landmark action was there. James Combs, a leading scholar of the period, highlights the long-term importance of Ned Ludd's revolution:

> The Luddite movement was one of the first flashpoints of a social conflict typical of modernity: that the traditions, arrangements and habits of the past must be sacrificed in the present for the promise of a more glorious future.[1]

Their cause weathered the original setback and came to life later. They were the early advocates of the trade union movement, which protected the English working class and became a major force in Britain's Labour Party.[2]

In the late twentieth-century a new branch of Ludditism surfaced in America, working often with newly empowered environmentalists, to oppose what Neil Postman called, in his influential book *Technopoly,* "the deification of technology." Such a cult is hard to resist, since it produces what is most cherished in today's society: big dividends.[3] Isn't Bill Gates widely touted as "the richest man in the world"?

Not all members of Gen X admire the rampant technology. James Combs collects what he calls popular fables of estrangement. They fill the music releases, and popular films such as *Mallrats, The Doom Generation, Dead Presidents,* and *Pump Up the Volume.*

They depict a new generation who have become "voyeurs of life aimlessly taking their pleasures but valuing nothing from the world that has chosen to abandon them."[4]

Will the result be a new generation of technophobes? Of neo-Luddites? These are questions that await answers.

Gardenspace to Cyberspace

The postmodern world has moved from gardenspace to cyberspace. As a result, I live in Cyburbia. You won't find it on the map. The best set of directions come in *Esquire* magazine (February 1996). There Randall Rothenberg coined the term for my town—Blacksburg, Virginia, its 36,000 inhabitants tucked away in the mountains, home of Virginia Tech. You may have seen the bumper sticker favored by our archrivals at the University of Virginia: "All dirt roads lead to Tech." Rothenberg asks if those dirt roads might also lead to a new digital utopia.

As someone here on the front line, on-line, I was startled by his provocative question. Was he really writing about our pastoral village nine miles off Interstate 81, far removed from city life with all its increasing inconveniences and hazards? For many over the years, Blacksburg would have seemed more like Thoreau's Walden Pond than Cyburbia. A friendly traditional southern town with Main Street and Norman Rockwell trimmings, it has affordable housing on large lots in town or in the nearby countryside. The state's only land grant college, Tech (also known as VPI, Virginia Polytechnic Institute) has plenty of land, some occupied by cattle, horses, and sheep. The gray limestone college buildings cluster around a grassy drill field, where student cadets, clad in Confederate gray, parade with heads held high. On football weekends, they drag an ancient cannon to the stadium, to be fired when our team scores a touchdown.

I accepted a job as professor in 1976, to set up new courses in the humanities and American studies. My wife and I bought "the house of our dreams" on our first trip to town, even before arranging to move. It was on a street with ancient shade trees, and a hundred-year-old elm in the front yard. The house, built by two retired Tech professors, was in the Georgian style, designed by the architect after whom the new architectural building at Tech was named. Our new home had Williamsburg antique bricks, a slate roof, copper pipes, oak floors, five bedrooms, and a spacious dining room with a chandelier imported from Europe just before World War II. There was a large garden in the back, an asparagus bed, and fruit trees. The garage, actually a renovated eighteenth-century log cabin with a shake-shingle roof, had a tool room and a root cellar. Both our neighbors, retired professors, were serious gardeners and fascinating local historians. Could it be that Blacksburg was about to be transformed, not into Paradise Found, but Cyburbia?

Everyone is aware, of course, that there has been rapid growth in the institute that locals still call "the college." The enrollment has tripled since 1970, and new colleges have been added. An aggressive administration has courted and welcomed the Electronic Revolution. We have a vast number of computers and a mainframe, which we later abandoned when Apple computers came into vogue. We concocted a new motto: "A University Putting Knowledge to Work," and got a whole bevy of new administrators. Teaching has been matched by great emphasis on publishing, and political correctness watches over the whole scene. We have changed with the times. But is Blacksburg really Cyburbia?

The change came without many of us realizing just what was happening. This was the problem: how to combine our quiet Virginia village and the new Electronic Revolution. Blacksburg Electronic Village (BEV) decided to take us backward in order to move us forward—they played the "down home card" first. "We'll make Blacksburg just the way it was B.F." (Before Ford—an abbreviation Aldous Huxley uses in *Brave New World*). Andrew Cohill, newly appointed director of BEV, promised that what they had in mind was to turn our bucolic village into a real-life community laboratory—had we had no "real life" ere now? Flooded by companies with twenty-first century information, Cohill insisted, we would be the envy of the world. This was no pipe dream—he could even tell us how the miracle would be financed.

To turn our village into a state-of-the-art laboratory would take big bucks—and Bell Atlantic was willing to invest millions to wire our village. Virginia Tech would contribute, and urge (even push) its faculty into the new age. Even the town of Blacksburg would help, giving out generous grants to local merchants and corporations that would get on the Web and in the swing. *Would it work?*

The same issue of *Esquire* in which we were christened Cyburbia described another phenomenon that links up with our electronic village. It's called BYC—Because You Can—and explains why we are lured into a technology that can be useless, trivial, and silly. Why do you resort to the computer rather than a simple (often quicker) phone call to Pizza Hut? Because you can. Why do you check in to find out what movie is showing, even when you have no intention of seeing it? Because you can. Why do you want to see how the surfing waves are coming in at Carlsbad, California? Because you can. Just get on the Web at http://sailfish.peregrine.com/surf/surf.html. Simple, huh?

I have another suggestion: let BYC also represent Because You Can't. Beware of Net traffic jams. The sexier the software, the slower the Net; the nearer the site, the harder to get on. Watch out for "flash crowds," following hot topics such as Princess Diana's demise. Promises, promises. An old Asia hand, I was surprised to get this invitation:

> Tired of Blacksburg? Take a virtual trek through Asia!

But only, the notice warned, if you are "ready to use Internet resources regularly." If so, you can "learn about the systems of government and politics in six countries, after a systematic introduction to each of these countries." You will also master "Asian regional dynamics and organizations." "Master" Asian dynamics sitting quietly in Blacksburg? Virtually impossible.

This question has been carefully researched by David Silver, a graduate student at the University of Maryland who wrote his MA thesis on Blacksburg's BEV. He follows the development of a scheme to create a community network that shares not only information and knowledge, but also goodwill. These things are easy to exhort, but hard to prove. Is the community helped if you order your pizza by computer, rather than by phone—or by going out, talking to neighbors, and picking up a pizza?

Will one major result be that more and more people will do their business, banking, and shopping at home? If so, Silver asks, is that the way to "build community spirit?"

Silver traces the public relations blitz promoting BEV. A well-crafted account appeared in European papers, such as Germany's *Frankfurter Algemeine Zeitung* (November 4, 1997). Published in German, but translated verbatim for an English dispatch, a typical passage reads:

> In all classrooms and dormitories there are monitors, thanks to the forward-thinking educators of Virginia who have decreed that all schoolchildren be computer literate by age eleven.

In addition to such optimistic statements, a series of pictures shows "happy people—young and old, students and townspeople—in a town blossoming more than ever." Somehow BEV sounds like another Disney World.

What is blossoming most, Silver suggests, are the many commercial interests featured prominently in BEV.

A high-minded "Statement of Our Vision" was prepared. BEV would pioneer the effort to set up a community network that would share information, knowledge, and goodwill. Of course, Bell Atlantic had another goal in mind—to test out new devices and systems that could be used around the nation and world.

So it was that in late 1994 the network took off. There were visible successes and extensive publicity. The media picked up the story. *The Los Angeles Times* called Blacksburg the most connected town in the world. *The Washington Post* also said Blacksburg was the most computer-connected town on the planet. That takes in a lot of territory.

Certainly there was plenty of opportunity to do business in BEV. You could call up accounts, advertising, marketing, wedding services, beermaking/winemaking, beekeeping, laundromats/cleaning, bird supplies, chimney sweeps, dance and driving instructions—the list goes on and on. There are more exotic opportunities: healing your emotions, changing your thinking, starting a village. In the early days there were even BEV links to two MUDs (multi-user domains). These allow users to "change names and personalities and set up their own chat spaces." Some people in Cyburbia decided to play the game. Log-ins using IDs "Prancer" and "Lusty" asked repeatedly, "Will you suck our dicks?" Another log-in, "Lunar," admitted he "Would like to lick Jell-O off of Elizabeth Dole."

With the new unlimited freedom of access and material, citizens in Cyburbia or anywhere else can depict a physical space that exists only as descriptive data. Masters of this new technique can simulate rapes in cyberspace. They can retreat to a place called Lambda Moo, a very large and very busy rustic chateau built entirely on words. There a character named Mr. Bungle, a fat, oleaginous man, wears a belt buckle inscribed: "Kiss me under this, Bitch!"

* * *

Rothenberg, sensing some of these long-range problems in his *Esquire* story, is not sold on Cyburbia. He has some dark thoughts on what is happening in Blacksburg and around the world. Like many others, he wonders if flashy short-term gains (such as have crowded the Internet) will divide us into small groups "as disconnected from the common culture as a schizophrenic is from the world outside his head." Is it good for a society, he asks, to be ripped into tens of thousands of little groups, each with its own agenda, its own mythology, and its own Web page? Could unexpected problems crop up even in Blacksburg?

You bet—Rothenberg gives an example. A Tech student posted four photos of nude women on a Web site for gay and bisexual men. With them was a message calling for homosexuals to "die a slow death." Should we arrange to cut out such offensive material or allow (as the Supreme Court has ruled) free access and free speech? How far can and will we democratize cyberspace? Who, if anyone, should regulate it?

Do the Wired Ones want to create, Rothenberg asks, a world without bosses or leaders, without cities or governments, and ultimately without moral or temporal authority? Is technology a force for good or a force for evil?

Some in newly named Cyburbia would push aside the dark shadows and answer "good." They would base their answer, in some instances, on efficiency and not morality: The new technology is magic, its achievements are breathtaking—silicon chips, instant everything, space walks, AWACS, medical breakthroughs, military marvels, the Internet, the World Wide Web—and on and on. The future is electronic. Tune in or drop out.

The new machines are logical and rational, if neither moral nor magic. Computers do not have brains or personalities—only a complex set of chips and wires that can distinguish between "0" and "1," and carry out commands. When one computer company brags that its products "have not only a brain but a soul," we know that hype can take over from hope. We must demythologize the computer.

Something of immense consequence is under way. The Blacksburg experiment, seen as microcosm, points to the much larger macrocosm. As yet, even naming it is uncertain: the Internet, the Web, Information Superhighway, global network, Infobahn, the Matrix, virtual reality, cyberspace, and so forth. The buzzword now linked with almost everything, is "cyber."

"Cyber" is spreading like a media virus, and turns up everywhere. Stacy Horn has written a new book called *Cyberville: Clicks, Culture, and the Creation of an Online Town* (1998). Like me, she sees such a creation as changing technology but not human nature—or local problems.

Norbert Wiener, in his classic 1948 work *Cybernetics*, coined the word "cybernetics" for the new automatic technology. At that time, there was no fixed name or image for "computer." The first one, a postwar construction at the University of Pennsylvania nicknamed ENIAC, was only a few years old. Wiener's name for it was "an ultra-rapid computing machine." He was euphoric about future possibilities.

But in 1950 he wrote *The Human Use of Human Beings*, which sounded a different note. He regarded information technology as a threat to short-term social stability, and perhaps as a permanent disaster. The inventor of cybernetics intended to act as its conscience, asking: What if cybernetics is left wholly in the control of short-sighted, profit-maximizing corporations? Nearly five decades later we can see why and what he feared.

Two years after Wiener issued his warning, Kurt Vonnegut Jr. published *Player Piano*, a tale of technocratic despotism. When machines extend their power over crafts and skills, they destroy what has given meaning to people's lives. The technicians may argue that machines are our slaves; but Vonnegut responds that anyone who competes with slaves becomes a slave.

Perhaps, as Michel Foucault and others suggest, we are moving into a new *episteme*—a new way of conceiving of knowledge itself. At the center of this new day is the computer—the latest and most spectacular tool for collecting and storing information. Certainly we have the information—but what of the knowledge, the wisdom? Do they exist on a different and higher plane? That question remains to be answered.

The story of BEV is still in progress, and no one knows what the outcome might be. The sponsors are anxious to protect their multimillion dollar investment as well as to light up the electronic sky. A new contract to build a smart road with electronic signals and checks from Blacksburg to Roanoke raises their hopes. They are attempting to change the old bumper sticker from "All dirt roads lead to Tech" to "All smart roads lead to Tech."

Virginia Tech made a dramatic announcement on August 28, 1997. Beginning with the fall semester, all incoming students were required to own computers. The stated justification is that for these students, computer literacy needs to be just like reading and writing—part of one's very nature. The long-range implications of such an idea are staggering.

Simultaneously Bell Atlantic announced it was giving a $250,000 grant to Virginia Tech's Institute for Connecting Science Research to the Classroom. More money, more wires. The next day some computer buffs met to celebrate what they called "the great leap forward." No historian was present to remind them that it was the very phrase Chairman Mao had used in China.

* * *

The year 1996 brought new problems for BEV. In March the subsidy for non-Virginia users was withdrawn. No more low $8.60 per month rates. They were forced to subscribe to a local Internet service provider, whose rates were much higher.

Feeling tricked and betrayed, once-happy Blacksburg supporters lashed out in local newspapers and talk shows. "The true goal has been to build a little empire by signing up as many people as possible in order to generate a bunch of positive PR," was an angry response when I made inquiries. "A classic 'bait and switch' scam," another complained. "As soon as it looks like it might work we'll turn you out in the cold." The Era of Good Feeling was ending.

Then things got worse—flaming set in.

Flaming—the use of vulgar language and references—was particularly disturbing to some of BEV's strongest supporters, such as senior citizens and church groups. They were confronted by shocking material on their screens; it need not be repeated here. When does freedom become license? That question hangs heavily over all of cyberspace. Thus does reality burst the bubble of your on-line Hometown Village, and prove that human nature can't be changed electronically. Flaming continues, and spreads throughout the land.

Ever since Marshall McLuhan electrified readers in the 1960s with his "global village" hyperbolic rhetoric and utopian idealism, the spread of what the French call "le McLuhanism" has received wide attention. In the past few years the use of "global collectivity" has been joined by "local connectivity," as panaceas for our endangered communities. BEV is one of a number of concrete responses to that idea.

The vision statements made by BEV—and many similar experiments in wired towns—stress two major purposes: to foster an electronic town square and to develop an on-line shopping mall. A series of computer workshops and seminars were conducted for Blacksburg citizens, students, and community groups. A printed and on-line publication called *About Town* sang the praises of the new scheme and won converts. The advance publicity was extensive.

The existence of multi-user domains would, it was hoped, allow the whole process to be shaped by users as well as administrators, assuming a democratic framework and creating a "real-life community laboratory." BEV administrators drafted their own goals and simply applied them to the network.

Because of the prominence of Virginia and the many people there who already made extensive use of computers, the Educational Center was given special attention. But here too private interests worked their way into the "public" realm of the BEV. A commercial educational company, Kaplan, was included and featured prominently. As it turned out, a large share of attention and space went to the shopping mall.

As time went by, it was clear that Virginia Tech was anxious to expand the electronic village idea. In 1997 the University's foundation bid $1.1 million in the FCC spectrum auction for four licenses that cover the Roanoke, Danville, Martinsville, and Johnson City, Tennessee areas. It was seen as a way "to bring high-speed Internet to rural and mountainous communities." This has yet to be achieved.

In April 1998, BEV announced "several new initiatives," including adding "port per pillow" connections with several apartment complexes in Blacksburg. BEV also wanted to implement a NAP (network access point) in the downtown area. This would help solve a current problem—an e-mail message sent from Virginia Tech to any of the other modem-pool providers typically makes an extended trip around the East Coast, including Washington, DC.

The difficulties of Blacksburg pale beside those encountered in another highly publicized electronic experiment, in Malaysia. Called "Cyberjaya" (meaning "cyber success"), this project set out to link every house, office, and shop in a new city hacked out of Malaysia's dense jungle. The planners, with the slogan "Brave and Bold," envisioned Cyberjaya as "somewhat like California's Silicon Valley," starting as a local interactive grid, then joining up with others worldwide.

Then came the unexpected and disastrous crash of the Malaysian economy in the closing months of 1997. Plans to construct "the world's longest building" ground to a halt. The question on everyone's lips was: Will Cyberjaya turn out to be the tiger of technology—or just another half-vacant industrial park?

Are we in fact destined to be a global village, in which all the small cyburbias are linked electronically? If so, what sort of world will it be?

From gardenspace to cyberspace is a jolting leap, not only in technology but in metaphysics. If you want to be happy forever, Voltaire said, plant a garden. Of course I can still do that. Presumably the corn and tomatoes will grow as in former days. What is different is the atmosphere, the community in which we live, the way the university functions, and how we relate to the rest of the world.

If we are indeed entering a newly electrified form of the cave that Plato urged us to leave, we must consider some of the problems that will result. It has already raised some of them with me, my family, students, and friends. We know (and are constantly told) what we have gained. But what have we lost? That question will not be quickly or easily answered.

ENDNOTE

Much of this chapter is a personal essay about my hometown. However, David Silver's 1996 unpublished master's thesis on Blacksburg Electronic Village in the American Studies Department at the University of Maryland supplied valuable statistics and information. For further information Silver may be reached via e-mail: googie@wam.umd.edu.

Two pertinent recent books with excellent bibliographies are:

1. Clifford Stoll. *Silicon Snake Oil: Second Thoughts on the Information Highway* (New York, Doubleday, 1995).
2. Gregory J. Rawlins. *Moths to the Flame: The Seductions of Computer Technology* (Cambridge, MA, MIT Press, 1996).

Two important books from the 1980s:

1. Peter Hall and Ann Markusen, editors. *Silicon Landscapes.* (New York, Routledge, Chapman and Hall, 1985).
2. Geoff Simons, *Silicon Shock: The Menace of the Computer Invasion.* (New York: Basil Blackwell, 1985).

PART III: THEMES

People

Popular culture is people's culture. The Greek word for people, *demos*, is the root of our word "democracy."[1] But not *all* people. The Helots in ancient Greece—nonpeople—had no rights or political voice. Democracy (*demos* the people + *kratos* authority) was, said Plato in the *Republic*, mob rule—the worst possible form of government. We are startled, taught from childhood that democracy is the best form of government. Doesn't our Constitution begin with the famous words, "We the people of the United States"? We seldom ask just how many of "the people" were consulted when the Constitution was written—or were allowed to vote when it became the law of the land. No one intended that *all* the people would vote. They never did and they still don't.

The *demos* have a long and complex history, beginning centuries before the Greeks and their insights. Earlier *demos* made Sumerian funeral effigies, built the pyramids, painted cave walls. Duplications and multiplication of sights and sounds took place when history was still blind. Popular anonymous arts stretch back to an unknown dawn. For millennia the chief medium (from Latin *medium*—intermediate, between) was the human voice. Eventually, when alphabets were conceived, we began to get bits and snatches of what came to be known as history. Here again the Greeks took the lead. Thucydides, who lived and wrote in the fifth century B.C., is considered the greatest historian of antiquity.

Thucydides and his Roman counterpart, Tacitus, assumed it was the nature of common people to be irresponsible and uninformed. Over the ages, leaders and rulers have feared the revolt of the masses—and have offered bread and circuses (now called "entertainment") to appease them. Even in the newly formed United States, committed to the notion that we are all born with inalienable rights, Alexander Hamilton did not trust the common man: "The people, Sir, is a beast." One wonders how Mr. Hamilton thought we could have a democracy without trusting the *demos*.

Most of us feel more comfortable with Lincoln than Hamilton. God must have loved the little people, Lincoln noted—he made so many of them.

What did the little people do, we wonder, in those untold centuries before we have any records or clues? Did hunter-gatherers, in their leisure, invent what we would now label popular culture? Was it not hunting and gathering, but sharing, that made us human? How did we get from Ape Man to Batman?[2]

We followed the pattern of the ages. Common culture and people were for thousands of years illiterate. The little we know about our distant ancestors comes from the recorded words of literate outsiders. Such men and women did and do find it difficult to comprehend people quite unlike themselves. It is impossible, George Dorsey writes in *The Story of Civilization: Man's Own Show*, for one set above his or her fellows to think clearly about social justice or to feel humanely. "If history shows anything, it is that power over people, like a habit-forming drug, leads to a lust for power."[3]

Those who rise from the ranks or win rebellions—Spartacus, Gracchus, Robespierre, Nat Turner, Castro—are not immune to that lust. Power has never been class-typed and it's always been popular.

That lust sometimes blinds one to the truth. To dismiss ordinary people as a mob, beast, or crowd does them a grave injustice. Crowds were and are composed of people with different needs, impulses, and goals. We are all too quick and glib in speaking of the urban poor, lower classes, underclass, or the *menu peuple*. Such phrases, George Rude points out, are often terms of convenience, or frank symbols of deep prejudices—not verifiable historical phenomena. Crowds, like history, are constantly changing and regrouping.[4] Who knows who will be in the next one? Turn on the nightly news and you may see your neighbors or children in the middle of the crowd.

Just when Europe's traditional oral-rural culture was being engulfed by the Industrial Revolution, scholars and intellectuals "discovered" the people and the folk. The leading proponent was Gottfried von Herder (1744-1803), who popularized the idea of *Volksgeist*—folk spirit. This spirit, he maintained, is the result of tradition, environment, and the times: tradition being the unique history of the people, environment the physical conditions under which they live, and the times (*Zeitgeist*) being the outlook of the moment. To prove his theory, Herder began to collect and publish *Volkslieder*—songs of the people—and *Volksmärchen*—folk tales. The practice spread from country to country—*folkviser* in Sweden, *canti populari* in Italy, *nepdalok* in Hungary, and *narodnye pensi* in Russia. Europe suddenly "discovered" the folk.

America, which had fought hard to abolish the hierarchies and class distinctions of Europe, did not begin to make highbrow versus lowbrow distinctions and research until the twentieth century (the "brow" terms were used by Van Wyck Brooks in 1910). A literary landmark was Gilbert Seldes's *Seven Lively Arts* (1924), which dared to examine and praise Charlie Chaplin, Krazy Kat, and the circus. No one at that time used the term popular culture, and librarians couldn't find it in their periodical indexes. In the 1950s John Kouwenhoven and Sigfried Giedion called attention to what they called the vernacular—the unselfconscious efforts of common people to create satisfying patterns, not inspired by tradition, but imposed by the driving energies of a new social structure. These two scholars did not have a large following. H. L. Mencken, Edmund Wilson, and Lewis Mumford were more successful, but they never thought of themselves as defenders of popular culture.

The Big Thaw came in the Kennedy Years, when the highly touted New Frontier opened new avenues of expression and research in an upbeat America. The heading "Popular Culture" appeared in the *Readers' Guide* for the first time in 1960.

With youth asserting that popular culture was their special domain and pop art flooding the galleries, the lines between pop and elite eroded. The Old Guard gasped when pop singers performed at the prestigious Metropolitan Opera House. Gilbert Seldes's ghost might have chuckled at a 1970 specially televised event at the Met titled "A Celebration of the Lively Arts."

Not all the new arts were lively—some were corny and contrived. Folklore was sometimes turned into fakelore, and this is still very true today. We get faint (if phony) folk echoes today in the United States with theme parks such as Dollywood, and Nashville's Grand Ole Opry.

Rather than condemn the pseudocowboys and hillbilly pretenders, strumming their electric guitars, we might ask what has given them this strong folk nostalgia—this idolization of an ephemeral past. Why do so-called urban folk singers dream of pink cherry blossoms and a sow that got the measles and died in the spring? As city crime grows and sirens screech, they remind us that "never did hoof of beast go down on a lark's nest." This may be why they join the zealous ecology crusade, greenpeace, and the frantic effort to save snail darters and hoot owls; and why they yearn to join Henry David Thoreau at Walden—if they can have electricity and modern plumbing.

Not only their environment but their food must be pure and uncontaminated. Down with animal fat, cholesterol, saturated fats of all kinds, additives, and preservatives! If it's not natural, don't eat it. How pure is pure?

Romanticism, the nineteenth-century movement that swept across Europe and America like wildfire, idealized the *demos*, finding in them virtues that the new urban industrial culture was ruthlessly destroying. England's great romantic poet, William Wordsworth, defined poetry as "man speaking to man," and wrote sentimental poems about poverty, hunger, and archaic ways. William Blake damned the "dark Satanic mills," and artists gave visual evidence. Handicrafts and rural isolation seemed somehow pure, renewing, almost holy. Getting and spending, Wordsworth warned us, we lay waste our powers. This notion keeps resurfacing, as in the counterculture of the 1960s and various movements of the 1980s.

Much less sentimental but very influential were the Socialists and Marxists, who also rediscovered the *demos* (renamed "proletariat") and made them the basis of an ideology. For Karl Marx, labor was wealth, and class revolution the necessary solution. If the workers of the world united, they had nothing to lose but their chains. Lenin turned Marxist thought into political and military action, and helped create a socialist U.S.S.R. that would give the world a new superpower and scheme for world domination.

After World War II, Soviet influence spread throughout Africa and Asia, and "Red scares" gripped Europe and America. But in 1989, the Berlin Wall (built to keep East Germany free from capitalist pollution) came tumbling down. Once more the people were left to clean up the mess. Nor was all that mess on the eastern side of the Berlin Wall.

So, as the twentieth century ends, the endless struggle between the *demos* and their "betters" (whatever name they choose to use) goes on. The common man—whoever that is—does not seem very attractive to those who consider themselves uncommon. James Thurber's trenchant quatrain in his book *Fables for Our Times and Famous Poems* summed it up well:

> I never saw a common man,
> I never hope to see one;
> But I can tell you anyhow
> I'd rather see than be one.

Others worked hard to depict the *demos* in a new light. A more sophisticated generation of Marxists, in the East and in the West, disproved long-held clichés and stereotypes. Clichés have no certifiable meaning; they are merely signals that suppress the word's history in depth, and send only a faddish clue. Instead of showing us reality, they conceal it.[5]

In England the Society for the Study of Labour History gave new interpretations of music halls, alehouses, working-class movements, and the rise of popular church music.[6] Deric Regin's *Culture and the Crowd: A Cultural History of the Proletarian Era* (1972) argues that the proletariats are no longer the factory workers, the slumdwellers in capitalistic cities, destined by a nineteenth-century formula to generate the violent overthrow of the bourgeois. Instead, they are today chiefly marked by being cut out ruthlessly from the vital functions of their own society. They are the aggregate of outsiders, fringe-dwellers; what the American novelist Ralph Ellison aptly called the *Invisible Man* (1952).[7]

The so-called Frankfurt School, exiled from Europe and writing in America, gave old Marxist ideas a new twist. Theodor Adorno, Max Horkheimer, and Herbert Marcuse were its leaders. They argued that the new woes of the industrial world were inherent in the development of the West. America was the archvillain in their view. For most people caught up in the high-tech world, the very idea of escape smacked of treason.[8]

More and more, proletarian culture was seen as the self-expression of mass consciousness and mass-mediated culture. This resulted in what some chose to call *anomie*—social or political apathy. Today's proletariat, watching the tube or supporting the mass rally, swears allegiance to his or her last security—being part of the crowd.

Others question this conclusion. Surely there has been some leveling of values in social consciousness and ethics—but does this mean a diminishing of self-consciousness? Has mass media changed the whole concept of *demos*? If so, how?[9] Contemporary history is beset with baffling questions.

Might not that well-entrenched cliché "The Age of the Masses" be a misnomer? Do the "masses" actually rule in Russia, China, or North Korea? Hardly. Do we err to think of the masses as being either good or bad? Are they not in fact neutral, like the media and machines that play such a role in their lives? The *demos* are what they are not because of intrinsic development or ideology, but because of the culture in which they must survive.[10] They are always in a state of flux.

Meanwhile, the culture of abundance has permeated America. Much of the old class analysis is meaningless. Leisure, recreation, and celebrity have replaced the old Puritan values of hard work and self-denial. Consumerism takes control; we have glut amid mass poverty, unchecked credit in the midst of unmanageable debt. The whole cultural matrix is altered.[11]

People long called common, primitive, or mass seem to be far more complex and sensitive than most intellectuals have wanted to admit. In the last generation this has been well demonstrated by structuralists such as Claude Lévi-Strauss and Jean Piaget; theorists such as Noam Chomsky and Northrop Frye; critics like Paul de Man and Geoffrey Hartman; social scientists such as Hayden White and Dell Humes. Age-old attitudes are being pushed aside and abandoned.

We are undergoing the change from a modern to a postmodern, an elite to a multiclass order; the process is still in the unfolding stages. It will change the name and nature of popular culture.

In *The Working Class Majority* (1974), Andrew Levison suggests that we are only beginning to understand workers. They are not more conservative or racist than other classes, and may not deal in clichés and stereotypes nearly as much as the college-educated middle class. If we would ask workers more appropriate questions, we might come up with some very startling answers.

We have known this for a long time. Walt Whitman suggested two crucial but still unanswered questions in 1855, included in the poem titled "By Blue Ontario's Shore":

> Are you really of the whole People?
> Have you vivified yourself from the maternity of these States?

Few if any can answer "Yes."

Lore

Lore: an ancient term reaching back into the dim past. An apt translation is "learning," applied especially to knowledge of a traditional or popular nature. The word itself is archaic, but linked with other words—as in folklore—it thrives today. All lore carries us beyond history into mythology, rituals, and legends—channels into which creative instinct and insight flow. Every historical epoch creates its own lore. Like Proteus in the *Odyssey*, Chapter 5, it takes "all manner of shapes of things that creep upon the earth, of water likewise, and of fierce fire." We shall examine some lore of our own time which may help to describe and interpret the world entering the lore-laden twenty-first century.

The very fact that "lore" is not one of the buzzwords, clichés, or sound bites usually employed gives it an advantage, a clarity, which might shed new light on my task; and help connect us to an essential past often forgotten or ignored.

What did "lore" mean when it was full of life and vigor? Teaching, wisdom, counsel; the whole body of knowledge possessed by a people, class, or nation. It stretched back in time to English *lar*, Dutch *leer*, and German *Lehre*. Geoffrey Chaucer wrote of "Christes loore" in 1386, and Hernyson used the term a century later.

In 1812 Jacob and Wilhelm Grimm published their famous *Kinder- und Hausmärchen*. In 1820 London's *Gentleman's Magazine* wanted to substitute the English "lore" for the Greek "ology"—using birdlore instead of ornithology and starlore instead of astrology, for example. The Greek prevailed.

"Folk," however the word is derived or defined, is of profound and primary importance, for it is the rock on which all cultures are built. The original meaning of the word is best preserved, according to the *Oxford English Dictionary*, in Old Norse—"a people, tribe, or race"—although its Lithuanian origin means "crowd." The word marches through English language and literature for centuries, from "folces" in *Beowulf* and "our folk" in John Wycliffe (1330-1384) to folk rock in the late twentieth century.

Generally the stress of the word is on the mass as opposed to the individual—the vulgar rather than the elite. "Folk" generally refers to a particular subservient class—retainers, followers, servants, workers. In our time the word has become chiefly colloquial, being superseded in more formal use by the word "people." Yet "folk" still crops up in unexpected places. On January 16, 1998 the Associated Press carried a major story connecting the "folk" to electronic telephone communications; the headline read "Baby Bells want Net folk to pay."

This headline reminds us that folklore keeps up not only with the folk, but the whole culture. Right now the "lore" centers on computers, the Internet, and the World Wide Web. The "folk" are not bedecked in coonskin hats waiting for a critter to crash through the underbrush. They are dressed in business suits, waiting for the electricity to come back on after their computer crashed. Poor folk!

Writers have continued to refer to the folk throughout the ages. The great English poet John Keats wrote these lines in "The Eve of Saint Mark":

> Twice holy was the Sabbath bell
> That called the folk to evening prayer.

Who hasn't heard renditions of Stephen Foster's "Old Folks at Home," and greeted close friends informally saying "Howdy, folks"? Millions of people have gone to the movies and watched film cartoons end with the tag line: "That's all, folks!" And the most famous small car of the twentieth century, throughout the world, is the Volkswagen Beetle.

Then there are the countless combinations, when folk is linked to lore, belief, custom, literature, name, song, or speech. Add to that what scholars call folk etymology—popular perversions in which words are linked, such as folk-free, folk-leasing, or folkstead.

In all these forms, "folk" signifies beliefs, legends, customs current among the common people, or folk. Words with the same connotations widely used by scholars and historians include serf, slave (derived from "slav"), peasant, and commoner. Another word crossed the Atlantic with the Scots and still flourishes in the Appalachians—hillbilly. Other terms for the mountain folk include red-neck, white trash, grit, clodhopper, and turdkicker. The popularity of "country western" and "country" music has skyrocketed as the "country" itself is eaten up by interstates, malls, and parking lots.

There is a more formal way to label what many would call "lower classes." They are "traditional," meaning they follow ancient ways and live by the lore of the land. A thumbnail sketch might read something like this:

> For people around the world for many centuries, the land was the lore and the lore was the land. They seldom ventured very far from hearth and home. There were no clocks or watches: the sun set the time and the seasons set the pace. The womb of life, the earth, was their Alpha and Omega. Far off was lord, king, emperor, or pope. They all lived in different worlds and answered only to God.

Most of the folk, being illiterate, left their mark not with quills but with plows and trowels. The basic human story is written on the face of the land. The appearance of pastures, mounds, gullies, furrows, and in our time slums, skyscrapers, junk heaps, and waste sites, pieced together and interpreted are more profound than all the books and manuscripts ever written. The landscape mirrors our lives.

Landscape means many different things. On a green hillside, the poet sees intimations of immortality; the engineer sees a new sewer line. A farmer translates an acre into so many bushels of corn; a soldier into so many foxholes; the oil driller into so many towers; the architect sees a place for a shopping mall. We all see what we are prepared to see—the mind trains the eye. It has trained the folk for countless centuries.

Primitive people are largely limited to the horizon. The Ancient World encompassed the inner sea—the Mediterranean—but most of the world lay beyond its ken. Land and sea were worshiped in epic, myth, and song. Although preoccupied with nature (as the ancient arts of Egypt, China, Persia, Turkey, and Greece show), these civilizations had few schemes for altering it. Nature had its own purpose and design (*telos* in Greek). You defied the gods of nature at your own peril. Watch out when Zeus gets mad.

The earliest metaphor for unexploited sacred nature was the garden. The people whose land and lore brought forth three of the major religions that still shape large portions of the world—Jewish, Christian, and Muslim—all began with the Garden of Eden. Scholars believe its actual location was southern Iraq. How ironic that the Garden should become a high-tech battleground during the Persian Gulf War.

If the words "folk" and "lore" stretch back into the dim past, the linking of these words in the English language was suggested at a specific time. The credit goes to the English scholar W. J. Thoms. In 1846 he coined the term "folklore," hoping to preserve "the traditional learning of the uncultured classes." To him the term was akin to what we have come later to call popular culture.

If the idea was Thoms', the reality behind it was very ancient, taking us back to the caves and the tombs. No lore predates that of the folk, however they were named or housed. Egyptian examples have come to us from the twenty-eighth century before Christ. Surely there were even older ones that never got recorded in preliterate times.

Indians living many centuries ago supplied the *Jatakas* and *Panchatantra*, Arabia *A Thousand and One Nights*, Greece a whole spectrum of epics and fables. What was not lost has been transformed or adapted. In his great study of *The Legend of Perseus* (1977), E. S. Hartland tries to find the origin of this folktale, but concludes that in the end this problem is insolvable.

Problems such as this interested W. J. Thoms. He was both a problem solver and a collector. The same can be said of collectors throughout Europe, like those in the *Volkskunde* movement in Germany. There was a general understanding that something important was being lost; that poets and writers had to help preserve it. People who create folklore seldom try to define it.

Folk culture, prior to the 1960s, was most generally thought of as something found in isolated groups with special distinctions. That isolation might be spatial, occupational, religious, racial, or linguistic. Now we see that folklore exists in all aspects of culture; for example, the folklore of politics, corporations, and computers. We have greatly expanded the field, and brought popular culture under the new expanded umbrella.

Folklore is crucial to preserving and understanding popular culture. In some ways they are Siamese twins. But neither term can be quickly defined. Still, five key words keep emerging: oral, transmission, tradition, survival, and communal. A central theme runs throughout the literature: the vitality of life doesn't come from the top to the bottom, but like a great tree, from the soil up.

Among all the people who set the course of folklore research, few were more influential than Andrew Lang. His article "Mythology and Folklore" in the 1874 issue of the *Fortnightly Review* set a trend which is still followed. Not philology but ethnology should be the foundation. Folklore, Lang argued, could be seen as survival. This tapped into the new Darwinism, and worked well with the ideas of Edward B. Tylor, who became known as the "Father of Modern Anthropology." His *Primitive Culture,* also drawing on Darwin's work, was published in 1871. To know the mind of early people, one should study contemporary peasants. The idea is still prevalent in popular thinking.

In America, folktales of the African Americans and Native Americans intermingled. Joel Chandler Harris recorded them as the Brer Rabbit stories. In his *Autobiography,* Theodore Roosevelt tells of hearing the Brer Rabbit stories from his mother and Aunt Anna.

Meanwhile Europeans worked on the problem of how to catalog folklore material. In 1910, Finland's Professor Antti Aarne developed a Type Index, now the standard system, allowing the storage and retrieval of folklore variants in a systematic archive. This historic-geographic method reduced variants to outlines of their plots, and devised distribution patterns. These may be assembled as archetypes. It was a giant step forward for folklore studies.

The essential procedures of the Finnish method were set forth by Kaarle Krohn in *Die Folkloristiche Arbeitsmethode* (1926) (The Folkloric Work Method). Leading practitioners of the method, besides the Finns, were the Norwegian folklorist Reidar Christiansen, the German Walter Anderson, and the Americans Archer Taylor and Stith Thompson. Thompson translated and expanded Aarne's Type Index in 1928 and developed the first *Motif-Index of Folk Literature* (published from 1932 to 1936), a huge catalogue of folk narrative elements that may variously combine to form whole folk narratives.

Much of this analysis has centered around that fraction of humankind known as "Western," and the much smaller one known as American. Most people are neither, of course. Who speaks for *demos* in Asia and Africa? One of the key experiments of the century has been in mainland China, where leaders aim to create a new culture; to transform the basic patterns and behavior of the world's most populated nation, using such "popular media" as folk songs, opera pictures, and wall newspapers as weapons. How well has it worked? We have only begun to look for answers.

If all this scholarly activity was helping to preserve and protect the people's legacy, it was not attracting much attention from the American people themselves. Their job is to create, not catalog. They took their hillbilly culture into Nashville, Atlanta, and New York. By the 1990s, country western had moved to the top of the electronic culture charts.

There is general agreement that folklore is the verbal art of the folk, transmitted by oral tradition. The folk, this implies, are those who can't, don't, or won't read. Those who can and do read have found any number of categories and definitions of folklore. The *Standard Dictionary of Folklore, Mythology, and Legend* lists twenty-one. All assert that folklore is most often found among isolated groups with unique histories and points of reference. This isolation may be spatial, occupational, religious, racial, or linguistic. It may be a combination of these as well. Usually the setting is rural and religious; behavior is traditional and seasonal; the sacred prevails over the secular.

There are still enclaves of genuine folklore in America: in mountain hollows, on ships and in prisons, among the Pennsylvania Dutch, Gullah Blacks, or French Canadians. All folklore is orally transmitted, but not all that is orally transmitted is folklore.

Folklore is highly volatile and adaptable. When early colonial settlers from England brought with them to the New World a folk song known as "The Three Ravens," they eventually transformed it into "The Three Crows." Even today, in isolated backwaters of the Appalachians, one might still hear the powerful and macabre tune:

> There were three crows sat on a tree
> As black as any crows could be;

One of these crows said to his mate
"What shall we do for food to eat?"
"There lies a horse in yonder lane
That has been only three days slain
We'll sit upon his bare back bone
And pick his eyes out one by one."

This stark and brutal folk rendition never found its way to Nashville's recording studios and the networks. The glue holding electronic America together isn't traditional folklore but a made-for-money substitute. It calls for a special name: fakelore.

Fakelore is pseudofolklore; the outer form without the inner substance, manufactured and peddled for profit. The transitional phase between traditional folklore and electronic imitations is one of the less admirable manifestations of popular culture. But it is alive and well, and deserves our attention.

When did folklore begin to come not from the folk but specific authors? We cannot be sure. Faking folklore is surely no new thing. The Grimm brothers found a good market for the fairytales they recorded. Poets have long imitated the ballad. James Macpherson "discovered" Ossian's folk poems in the eighteenth century and Heinrich Heine, William Wordsworth, and Christina Rossetti all "borrowed" from the folk idiom and made it their own.

After 1860, the homespun yarn never again became nationally popular. A century later literary scholars and folklorists would unearth specimens as mere curiosities. The steady destruction of the bases of folk culture attuned to nature and rural life continued in the twentieth century. "Folk," H. W. Fowler wrote in his 1926 *Dictionary of Modern English Usage*, "has passed out of the language of the ordinary educated person, so far as he talks unaffectedly."

Fakelore

The word "fake" is of obscure origin, although it is much used today. "Fake" has come to mean counterfeit or imitation; something "doctored up to deceive." A faker is one who fakes—a con artist you may expect to meet at the fair, the festival, or the street market. Beware! All that glitters isn't gold.

The term fakelore seems to have come from academia, meaning phony or fake folklore. Several of us thought we had invented the term—none so vigorously as Richard Dorson, who denounced anyone who "used his word" without permission. During the 1960s and 1970s he was a kind of "folklore czar" at Indiana University. At his death in 1981 his reign ended. An admirable folklore program still flourishes at Indiana University. Professor Henry Glassie, whose books on Ireland, Turkey, and Bangladesh have reinvigorated the whole field, teaches in the program.

Whoever invented "fakelore," fakers have been around for centuries. We have had fake relics, antiques, documents (the famous "Donation of Peter," for example), epic poems, paintings, and "literary discoveries and borrowings." Poets and songwriters have long taken what they need from others. Every culture has its share of pseudofolktales and epics. But the pace has quickened in our century as the new mass media invented new forms of fakery. The dynamics of transmission changed. A central spot for double talk, sound bites, and fakelore in the late twentieth century is Washington, where fakelore invaded the Congress, White House, and Beltway.

In the hinterlands we have constructed "theme parks" that take us "back in time" to fake places—as at Busch Gardens, Old Tucson, or Disney World. Instead of real jungle beasts, one can go to Disney's Wild Kingdom in Orlando and watch plastic elephants shoot water out of their trunks, while a giant plastic crocodile opens his mouth and waits for awed spectators to throw coins in it for good luck. This might not be fakery—a better term might be ersatz amusement.

We have convincing fakes that fooled thousands and made millions. In 1869 workmen in the little town of Cardiff, New York, "discovered" a buried petrified figure over ten feet tall, weighing 2,990 pounds. The Cardiff Giant took the nation's breath away. Oliver Wendell Holmes found the giant "of wonderful anatomical development," and Ralph Waldo Emerson called it "beyond my depth, very wonderful and undoubtedly ancient." Whatever it was, Phineas T. Barnum knew a good thing when he saw it. He quickly offered $60,000 for three-month's rental.

Eventually the truth emerged. George Hull had had the giant carved in Iowa, brought secretly to Cardiff, and buried for a prearranged "discovery." You can still see the Cardiff Giant at the Farmer's Museum in Cooperstown, New York. I took my children there and they were convinced it was real.

The Gilded Age was a golden age of fakery. Science fiction novels such as H. G. Wells' *War of the Worlds* (1898) and Jules Verne's *20,000 Leagues Under the Sea* (1870) sent shivers up the world's spine. In 1896, thousands of U.S. citizens saw cigar-shaped airships overhead. Investigating this, Robert Bartholomew decided it was a case of "collectively shared wish fulfillment resulting from rapid technological strains." Sure enough, a few years later the Wright Brothers actually managed to fly from the sands of Kitty Hawk.

There was nothing ethereal about Buffalo Bill's Wild West Show, in which tame Indians bit the dust. Cody staged fake shoot-outs and clashes. (Sometimes Cody drank too much. During a performance in Denver he almost shot the mayor's wife.)

Writers such as Frank Murdock were getting rich from fakelore. His *Davy Crockett* (1872) has Davy trapped in a cabin during a blizzard, while the heroine, Eleanor, reads him the story of young Lochinvar. Only "the strong arm of a backwoodsman" kept the wolves out. All works out well, and Eleanor is "happy in heart and home of Davy Crockett." Compare all this with the gritty prose in the early Crockett almanacs. Or read the sugar-sweet prose in David Belasco's *The Girl of the Golden West* (1905), used as the libretto in

Puccini's opera of that name. The blood and guts of the real Old West are gone. All that remains is golden.

After World War I, fake demigods burst upon the scene—Paul Bunyan, Pecos Bill, Big Mose, Strap Buckner, Annie Christmas, Bowleg Bill, Tony Beaver, Gib Morgan, Daddy Joe, and Whiskey Jack leading the pack. Their names and occupations vary, but they are all cut from the same bolt of cloth. They are part of the big build-up of the American ego. Where did this new breed of folk-fake characters come from? And what did their sudden appearance and success say about our changing popular culture?

* * *

A plethora of phony folk heroes and villains flooded our books, movies, and TV screens. *Life* published a "remarkable new series" (starting in August 1959) on "The Folklore of America," which was mainly fakelore. Pecos Bill didn't come from the Wild West, but from author Edward "Tex" O'Reilly. Annie Christmas's spicy saga can be traced not to the New Orleans brothels but to the typewriter of Lyle Saxon. Margaret Montague dreamed up Tony Beaver. Daddy Joe was contrived by Stewart Holbrook, Big Mose by Herbert Asbury, Whiskey Jack by Charles Brown, and Strap Buckner by Florence Barns. Jeremiah Jones refined fakelore by combining the appeal of sailors and cowboys—his Bowleg Bill specialized in riding giant tunas. "Maybe the scholars have been following a false lead," Bernard De Voto wrote in "The Easy Chair" (1955). "Maybe popular literature isn't a folk art at all." By then factory-made folklore had blazed a trail across American journalism, advertising, and entertainment like a jet plane racing across a cloudless blue sky.

One of the first writers to deal with the consequence of this was Walter Lippmann. As early as 1922, in *Public Opinion*, he gave examples of "oversimplified patterns" in American life that gave us ways to defend our prejudices by seeming to give definitiveness and consistency to our daily experience. For years thereafter Lippmann pointed to pseudoevents (the fakelore of journalism) that "ignore the world outside and concentrate on the pictures in our heads." In so doing, Lippmann piqued most of the master image makers. It remained for Daniel Boorstin, in *The Image: Or What*

Happened to the American Dream (1962), to dissect a pseudoevent: (1) it is not spontaneous, but planned and planted; (2) it is planted to be reported and reproduced; and (3) it is tied in with a self-fulfilling prophecy.

Fakelore is to folklore what the pseudoevent is to the real event. And thus the emergence, in the Eisenhower years, of a new Gresham's law of American public life: counterfeit happenings will always drive spontaneous happenings out of circulation. Poison tastes so sweet that it spoils our appetite for plain food. When the gods want to punish us, they make us believe our own advertising.

Perhaps fakelore in the grim days of the Great Depression was a form of escapism. The most famous example occurred on Sunday evening, October 30, 1938, when CBS carried Orson Welles' version of H. G. Wells' story, *The War of the Worlds*. It has become the best-known example of electronic fakelore and mass hysteria in the twentieth century. Following the broadcast, reports of suicides, heart attacks, and group stampedes flooded the press. Even today the whole episode is frequently recalled and revivified.

Much of the escapism of that decade took a much lighter turn—such as Fred Astaire's dancing in his top hat and tails, Shirley Temple sailing off on the good ship Lollipop, and Judy Garland skipping along the yellow brick road to see the Wizard of Oz. Then, eight million American servicemen went to war, inspired by Walt Disney's training films, flying bravely into the wild blue yonder. Folk, fake, and global warfare coalesced.

* * *

Both folklore and fakelore keep up with the times. In postwar America, UFOs, extraterrestrial creatures, and paranormal phenomena became front-page news. On January 24, 1947, Kenneth Arnold set off the "Saucer Hysteria," having seen "nine tailless and wingless objects headed toward Mount Rainier, traveling 1,700 mph." During the next few weeks, hundreds of sightings were reported all over America.

Nocturnal lights, daylight disks, and close encounters became buzzwords. A new science—Ufology—was born and a new "taxonomy of hysteria/delusion" put forth by scholars. Humans seem capable of creating episodes that are collectively shared delusions. Is this an advanced form of fakelore?

Whatever it is, ufology thrives inside popular culture, and in numerous books, movies, television shows, and documentaries. Examples are television series such as *The X-Files*, *The Psi Factor*, *The Extraordinary*, and *Strange Universe*; various *Alien* movies; eight *Star Trek* movies; and hundreds of "factual" books about aliens and abductions. The term "cultural folklore delusion" turns up at scholarly meetings. In the real world, people keep seeing UFOs, spaceships, and things that go bump in the night. The millennium is doubling the sightings.

Many of these "strange events" are deliberate hoaxes, involving still photography, video, balloons, and kites. Others turn out to be caused by meteors, weather balloons, or various aircraft. We know that mirages are created when strata of air come together and are compressed. These layers act like a two-sided mirror, and make deception seem real. There is also what experts call the "false memory syndrome." Some people really do believe they were kidnapped by aliens, as happens in science fiction.

UFO cults made front-page news again when thirty-nine members of the Heaven's Gate cult committed suicide in 1997. When the Hale-Bopp comet was discovered, cult founder Applewhite claimed that a UFO was hiding behind it. Their spacecraft would take members of the Heaven's Gate cult back to their world, once they shed their human bodies by killing themselves. That is exactly what most cult members did.

What is the future of fakelore? Who can say? But that it always has been, and always will be with us, seems certain.

Poplore

Defining poplore brings us back to an earlier dilemma—just what is pop, how can we reconcile its multiple meanings, and what role does it play in our lives? Like electricity—on which poplore depends—it is powerful, invisible, omnipresent. Since it eliminates old boundaries and parameters, it cannot be put in a neat box. One of the first persons to understand and elaborate on this was Marshall McLuhan.

He noted that in the twentieth century, folksingers ran in out of the noonday sun into the air-conditioned comfort of recording studios and nightclubs. In that process, they did more than fake the old style—they invented a new one. Call it popstyle.

This is why, in the 1990s, one can go to Korean or Japanese nightclubs and hear young performers who don't speak English lip-synch American country-western singers, and compete in popular Elvis look-alike contests for large sums of money.

Early collectors of genuine folk songs, such as Cecil Sharp, described them as austere, high-pitched, rigidly performed, turned into something charming, sentimental, sweet. To put it more bluntly, the guts were removed from folk music. Alan Lomax, a leading folk song collector, noticed that when Aunt Molly Jackson sang to country folks in Oklahoma, they called it the most beautiful music they had ever heard. When she repeated the program for city children, they compared her singing to a cat's yowling. To the first audience, the folk "gapped scale" and modal intervals were still familiar; for the second, raised on mass media kitsch, one had to adjust the intervals, remove the harshness, and electrify the guitar.

But the music and lore of the Elvis Presley era dwelt in a strange no-man's land. It aped and honored the old folk idiom and metaphor, like the "colonial home" in suburbia which had a fireplace, spinning wheel, and picket fence—complete with electric stove, sewing machine, and television. No one summed up this cultural schizophrenia better than Marshall McLuhan in *The Medium Is the Message* (1962):

> The past went that-a-way. When faced with a new situation, we tend always to attach ourselves to the objects, to the flavor of the most recent past. We look at the present through a rear-view mirror. We march backwards into the future. Suburbia lives imaginatively in Bonanzaland.

What McLuhan had done was study folklore, discern and reject fakelore, then discover poplore. He had, in a single lifetime, gone through the whole cycle I am trying to describe and document.

The new poplore is not the antithesis of traditional folklore. Pop is not slick but savage; in musical terms it avoids the chromatic scales of the nightclub and uses the old Greek musical modes. Both folklore and poplore avoid sentimentality, reject the "arty" approach for the earthy one, draw from primary materials, colors, and emotions. The line of force connects more with the stomach than the cerebellum.

Poplore may lack both the vitality and significance I attribute to it—only time will tell. But from the information available at this moment, I believe that Warhol, Rosenquist, Oldenburg, and Segal changed the course of American art. We will continue to enjoy Molière, Shakespeare, and Ibsen; but now that we have witnessed "happenings," the theater will never be the same. Playwrights can no longer work in isolation, since they confront a group on either side of their message. When performer and audience are wed, who can put them asunder?

And who can feel comfortable with terms such as Xeroxlore, e-mail lore, CD-rom lore, televisionlore, computerlore? If lore means learning, why not? Some of us know all too well how hard it is to "lore" a new super-size computer. Perhaps because the learning involved seems largely technical and mechanical?

Learning is never pure and unencumbered. We marvel at the scrimshaw carved from whale teeth by sailing ship crews. Yet they often got their subjects from advertisements—scrimshaw capitalism. They copied what was already there.

Consider the 1997 blockbuster movie, *Titanic*—a piece of popular culture triumph, netting millions of dollars. With a $200 million budget, it was a technological triumph—about a technological disaster, the maiden voyage of an "unsinkable" luxury liner that sank. The movie was in itself a hymn of praise to failed technology.

The performing arts have found a way to be involved in modern life, and to speak its idiom. Having been dominated by classical forms and European models for decades, they broke out in the 1960s, as traditional lines and barriers were pushed aside. Instead of using scripts, actors ad-libbed. A kaleidoscopic variety of subforms sprang up in the arts, and the street itself became a museum. This was the world of poplore. Some of it was artificial and pretentious, and passed quickly from the scene. But like every lore, it left its mark. Probably only 10 percent of it will survive. But can we ignore gold nuggets because they are found amidst tons of fool's gold?

Power plus structure equals a life of being. The poplore which sprang to life was pointing to something much larger than publicity—perhaps to a new ontology. We are much too close to make any final judgment. We can only record some manifestations, and let history take over:

Folk	**Fake**	**Pop**
oral	verbal	multisensory
traditional	nostalgic	experimental
realistic	romantic	psychedelic
earthy	sticky	tart
homespun	factory-spun	polyester
continuity	transition	explosion
improvised	ersatz	electronic
cowboy	Buffalo Bill	Bonanzaland
community sing	folk festival	Disneyland

New folklore springs up with new technologies. We will write about computer and Internet folklore, and many other recent innovations. Hackers are our new Robin Hoods, and they will have a "lore" of their own. Never sell short the folk imagination.

Fakelore obviously answers the needs of millions, and virtual reality may be all the reality some people seek and need. Perhaps it does leave them in Plato's cave, watching the shadows on the wall instead of the real objects. Suppose they prefer this to being dragged out into what seems to them a harsh reality? Do we have a right to drag them out?

Poplore seems to me a promising alternative for the new century. Can it somehow combine the better elements of the folk and fake worlds, and create something quite new? This is what pop art attempted—and in some cases achieved—in the second half of the twentieth century. Most of us shudder at Andy Warhol's comment that he wanted to be a machine. What kind of machine? Who will be in charge? How much of our humanity will be intact? These are questions that poplore must address. Who knows what the answers will be?

Our lore—our learning—is what makes us human. That is our divine gift.

Springlore

Spring—a word that leaps up, gurgles, flows. Derived from Old English *springan* and German *springen*, it is widely used both as verb and noun. We shall combine it here to form "springlore"—the study of what we know, think, and write about flowing waters, springs, and spas, and how they help illuminate popular culture.

For centuries, springs and watering places have served as healing places, stage sets, touchstones, cyphers, icons. Many a person has, like Narcissus, become transfixed by water.

Speak of the origin of water and you go back billions of years. We emerged from the water. When the astronauts look down on our planet, they see a harmony with oceans, rivers, and springs. They make our lives possible. We can live eighty days without food—only ten without water. Life, history, and water are all of a piece.

How do springs—and hence springlore—originate? The question has intrigued us for centuries. Old Testament prophets thought of water as the whip of vengeance of an angry God—hence the sagas of Jonah and Noah.

There were early watering places throughout all of Africa and Asia. Ancient Japanese storytellers reveled in springlore. They thought springs were guarded by *kappa*, curious creatures with webbed feet, duckbills, and concave heads.

Plato thought all waters under the earth were arranged in a circle, water flowing from one cavity to another—perhaps the first recorded example of both springlore and folklore? He further believed that waters have "neither top nor foundation, which causes them to float at the top and at the bottom." His star pupil, Aristotle, who began many of the studies we now label science, had a different idea. He thought springs and rivers are generated by condensed air. Herodotus and Homer both wrote extensively about water. Thermopylae, one of the best-known spots in Greek history, was the site of a famous spring.

The Romans were masters of aqueducts and springlore. Ancient Rome could boast of 800 public baths, some of which accommodated thousands of bathers. All things, Seneca believed, are in all things. Air changes into water, but it is never without water. Pliny writes in his *Natural History* that all waters are gifts of the earth. The saga of Europe's baths and spas (Spa is an actual resort city in Belgium, the name later becoming a generic word) is deeply rooted in European culture.

Springs and springlore flourished with Native Americans in both North and South America. At Easter time in 1513, Ponce de León landed in Florida, seeking the fountain of eternal youth. People have been flocking to Florida ever since, continuing the search.

In 1674 Pierre Perrault published *On the Origins of Springs*, launching the modern scientific inquiry that became known as hydrology, which is still popular three centuries later. By then pilgrims from old to New England were eagerly going to Lynn Spring, and William Penn was discovering mineral waters "which operate—like Barnet and Bath" in Pennsylvania. A 1698 London book dealing with springlore lists "ten springs in the British colonies," which were "all out as good as Epsom."

The English brought both their folklore and springlore to the New World with them. Who can say where English lore stops and Virginia lore starts? Not I. Roman baths were the model for Bath, England; Virginians established a "Bath" in Virginia—you can go there today (the new name is Berkeley Springs, in West Virginia) and see George Washington's bathtub. The bathhouse still has a large sign reading: ROMAN BATHS, thus acknowledging both the Roman and British heritage. In 1776 the General Assembly of Virginia decreed "fifty acres of land adjoining the said spring to be named Bath." Washington had known the spot earlier, when as a youth of sixteen he wrote in his diary: "We this day called to see ye famous Warm Springs. We camped in the field this night."

In 1783 there was a lively theater company at Bath. The local newspaper made the most of it, writing: "The muses follow freedom. Let us hail their residence here with us."

Bath and its inherent springlore flourished. "Hundreds annually come thither from all quarters," John J. Moorman wrote in an 1857 letter, "where days and nights were spent fiddling, faddling, and bathing." He also noted that gaming "is carried to a great excess."

The queen of the springs, however, was farther west, on the "Great Buffalo Trail," used since prehistoric times, traversed first by buffalo, then by frontier families moving west. The springs later came to be known as White Sulphur Springs. How ancient are the springs? Apparently they were known to the dinosaurs. The first dinosaur bones in the area were discovered by Thomas Jefferson, and named after him.

Many presidents followed him here. The President's Cottage, now a museum, was occupied by Presidents Tyler, Fillmore, Pierce, and Buchanan, and became known as the "Little White House." Thus did springlore and political lore join hands. Other assignments awaited the famous spot. During World War II a major army hospital took shape; and during the cold war, secret quarters were carved in the deep limestone rock so that Congress could hide out here in case of an atomic attack on Washington. Now that the cold war is over, tourists can visit the hidden quarters and be glad they were never needed.

In a world some choose to call a global village, local and regional values persist and even grow. Springs all over the world prosper in the 1990s.

Water is life, and the common history we share in war and peace—popular culture—is full of droughts, floods, deserts, drownings, bombings, sailings, and swamps. Water meanders into mythology. Rituals, legends, and fantasies bubble up. In some places springs are called eyes—they are round, clear, and deep. Folklore comes from the soil, springlore from the water. Springlore has a permanent place in our lores and lives. Oceans, lakes, beaches, and spas have a never-ending charm and attraction.

Captain Marryat summed it up in his 1839 *Diary of America*. Watering places all over the world are alike, he concluded. They are part and parcel of the human condition.

Icons

The war of the icons has long been under way.

Marshall McLuhan

Icons have always been with us, from cavespace to cyberspace. Every age and culture is iconic. We see the change when history becomes mythology, mythology begets ritual, rituals demand icons. Concepts end up as creeds and icons. This process is at work in Plato's Ideas, Christ's parables, Kant's Categories, Jung's archetypes, McLuhan's media, Gates' Microsoft. It shapes the careers of great thinkers and leaders as different as Buddha, Christ, Mohammed, Jefferson, Marx, Lenin, and Einstein.

Science has become our sacred cow, and has in many ways replaced religion. Science "went iconic" with Heisenberg and Bohr, and entered the academy as quantum physics, but made its impact on ordinary people when the atomic bomb exploded over Hiroshima. The bomb became the icon. Now it appears in the nuclear proliferation arguments, when nations such as India and Pakistan determine to make this powerful icon their own.

The word, from the Greek *eikon*, means an image, picture or representation that externalizes internal convictions. In Genesis, the Bible quotes God as saying, "Let us make man in our image, after our likeness" (1:26). This is done—"When God created man, he made him in the likeness of God" (5:1). Icons objectify deep mythic structures of reality, becoming tangible and even sacred. They are everyday things that make every day meaningful.

Icons are also cultural ciphers. They help us to decipher or unlock the mystery and wonder of the world around us. We can look at them objectively, but those who believe in them react emotionally—on the level of love and reverence.

Icons accumulate and alter—but also lose—meaning. The iconic Virgin Mary may not speak to the twenty-first century as she did to the thirteenth. The Nazi swastika does not motivate German youth of the 1990s, except for skinheads, as it did many in the 1940s. The British navy no longer rules the waves. The image precedes the idea in the development of human consciousness. The idea drives the image on to glory or to oblivion, as popular taste makes itself known.

Thinking in pictures is the first stage of icon making. The early cave paintings we have examined bear this out. They show that the mind is not so much a debating society as a picture gallery. We look with our eyes, see with our minds, make with our hands. Form and formula fuse. The word becomes flesh, and dwells among us.

Icons reach back to the dawn of time. Discoveries made by Heinrich Schliemann, who excavated ancient Troy, document the development from the formless stone to the human figure. Other findings take us back as far as 50,000 years, the best-known example being the Willendorf Venus, coming from the Aurignacian culture. What we now call female obesity, with exaggerated lower parts of the body, dominates certain female statues for centuries, but with meanings we can only guess at.

Early icons are full of vitality and power. They fascinated Greek philosophers, especially the Stoics.

As icons advance and mature they tie in with myth, legend, values, symbols, mindmarks, and idols. Religion puts great stress on icons, and the word often applies to religious images painted on a wooden panel. Now the meaning has greatly expanded, as we confront icons in supermarkets, gay bars, the Internet, discotheques, used-car lots, and funeral parlors. They pop up on billboards, bumper stickers, TV commercials, computer screens, and magazine covers. Heroes and icons survive and become indispensable: warm for devotion, cool for companionship.

Human nature remains a constant, and so do some icons. We must have icons to "look up to" and cherish. The ancient Egyptians filled their tombs with such icons, and the Romans their forums. Great religions thrive on icons, such as those used for prayer in the early days of Christianity. "I have seen a great many portraits of the Savior, of Peter and of Paul, which are well preserved," wrote Eusebius, Bishop of Caesarea (265-340 A.D.). The early catacombs, and many Christian churches ever since, abound with icons.

Yet the battle over religious icons has raged for centuries, especially after Islam prohibited them. Leo III, Emperor of the Eastern Roman Empire (717-740), tried to abolish the image of Christ in the church. The resulting schism was only partially resolved in the Council of Nicaea (787), which ruled that "the honor which is paid to the image passes on to that which the image represents, and he who shows reverence to the subject contained in it." A new cluster of words came into use: legend, belief, sacred object, relic, and veneration. With every new saint, every new miracle, new iconographic possibilities emerged. A communion cup was recently found in a remote Mexican village inscribed to Marilyn Monroe.

All sacred things must have their place. Being in place is what makes them sacred. Where but in the crypt of St. Peter's in Rome should we look for the bones of Saint Peter? Where should one expect to walk along Old Broadway if not in New York City, or find the Rat Pack if not in Las Vegas?

As our century ends we have reached "post time"—postmodern, post-Freudian, post-Marxian, postpatriarchal—but not post-icon. Today, as in every epoch, men and women need to make sense out of the universe. Even natural cycles of birth, growth, aging, and death create new cultural fads and fashions. Some of these become iconic. The paper-thin female body on the fronts of our fashion magazines—so unlike those of people with whom we live and move and have our being—have become iconic. So are some of the bone-bruisers who make millions by sacking quarterbacks or crippling tight ends. The meaning of new icons is profound and obviously falls in the realm of popular culture.

In an age caught up in both popular and mass culture, meaning and emotional response are still individual matters. Only as individuals do we know what we really think and feel. No mass media or external force can erase internal conviction, or what we learn in our formative years. Although subject to public pressures later on, we all maintain our private icon banks. We make regular deposits, and withdraw more than we know. We tuck away our private icons (notes, photos, trinkets) in the corners of drawers, and the corners of our minds, knowing we can draw interest from our deposits. If we can't do this, we quickly go bankrupt.

Toys are among the most precious possessions of people everywhere, in every age, often becoming iconic. They cross all natural and national barriers, helping us to journey to a magic world, and thus to control our behavior—the boy bounces his ball, the girl hugs her doll. Toys of infancy are of a fundamental and universal nature. Nothing is more prevalent than toys reflecting their people and period, and becoming icons. With and through them we pass from the dream world into the real world.

The symbolic value of a ball, fire engine, doll, sled, or electric train can stay with us all our lives. Citizen Kane's last word in Orson Welles' great movie by that name, was the name of his childhood sled—Rosebud.

We think of toys as being American or Western only because we know so little of their history. Almost everything found in our toy shops dates back many centuries—though with cultural differences. The yo-yo, for example, was long popular in China, and was imported to Europe in the eighteenth century. Our turn came later. Icons need no passports to move from culture to culture.

They come in all sizes, shapes, and materials. They can surprise us with an entirely new shape—like Russia's, which on October 4, 1957 launched the 184-pound device called Sputnik—the beep heard round the world.

Icons may be as small as a pill, as large as a skyscraper, as lethal as a nuclear bomb. They can be ordinary things with extraordinary power—telephones, cameras, computers, disks, cathode tubes, and T-shirts: all popular icons. Manna may still come from heaven, but much of our information comes through television, offering an iconic view of reality.

* * *

Icons and symbols, like other fully functional parts of popular culture, tend to be assumed rather than defined or recognized by ordinary people. We are so used to making a call on the telephone, and seeing the bell symbol for the Bell System, that we forget how it revolutionized our lives. This irresistible intruder in time and place weaves a seamless web throughout the world and awakens us at any hour, day or night. Computer zealots predict a similar web—but it is not as yet nearly so complete or effective as the telephone.

In the developed world, whoever and wherever you are, the Bell tolls for thee.

The emblem itself, designed by Angus Hibbard in 1889, went through various modifications, until in 1969 no words were necessary to convey its symbolic and iconic meaning. The switchover to the wordless icon, involving vehicles, uniforms, signs, ads, letterheads, and hard hats, took five years. Just how many life-and-death decisions hummed over Bell's wires in those years, no one can even guess.

The most popular American icon, now the world's, is the "classic" Coca-Cola bottle—possibly the best-known artifact in the world. Expendable, inexpensive, durable, and curvaceous, it has spawned an extensive lore of anecdotes, jokes, and behavior traits, creating a new "folk" tradition. Having seen and held Coke bottles in our hands for much of our lives, we have never thought of Coke as patent number 105,529, applied for on March 24, 1937. Behind the bottle is one of the most remarkable stories in the history of popular culture.

After a humble beginning, Coke has made "The Pause That Refreshes" a ritual occurring in 190 countries over 200 million times a day. The Coke Break cuts across all nations, classes, and races; Coca-Cola's annual sales surpass the entire economy in many of the countries in which it's bottled and sold.

The story begins with a morphine-addicted Georgia pharmacist, John S. Pemberton. In 1886, he combined Peruvian coca leaves, African kola nuts, and French claret to make a mind-numbing tonic. Like thousands of patent medicines of the time, his was attacked and modified. The wine was removed, as were traces of cocaine later on, and the beverage was marketed in soda fountains under a new name: Coca-Cola. The company was incorporated in 1892, and Coca-Cola was patented a year later.

First bottled in 1894, it was sold in straight-sided bottles until 1915. In that year Alexander Samuelson was commissioned to design a new bottle. We got the one we have known ever since. The renewed patent model of the bottle dates from 1937.

The metaphors used to describe the new design tend to be dated—"hobble skirt," referring to a dress design popular when the bottle was introduced; "Mae West," after the extremely curvaceous movie vamp, and "hourglass." One also finds references to the bottle's mouth, lip, and neck. In all these instances, the tendency to attach human attributes to the glass bottle is important. Competitors began using flashy, colored bottles and new lettering, cutting into Coca-Cola's sales. A modification was the obvious solution. (Henry Ford would have just the same problem with his own icon, when he insisted on producing only black Model Ts, while other companies used different colors and took away Ford's customers.)

In 1955, partially painted Coca-Cola bottles appeared, providing greater visibility, with a white label against the brown beverage. In the next ten years six variant Coca-Cola bottles appeared. How could the old workhorse bottle be maintained with enough variety to attract a new generation? Could the love of the old be linked to attraction for the new? The answer turned out to be Variant K, which has been in production since 1966.

Knowing this, my friend and colleague Craig Gilborn at Winterthur Museum in Wilmington, Delaware, proposed a scholarly study of pop iconology, centering on the evolving Coke bottle. How does the visual impact of an object affect our reaction to it? Can we measure nonverbal knowledge as we regularly measure and judge verbal knowledge? Professor Gilborn reminds us that Samuel Johnson notes in the preface to his 1755 *Dictionary* that words are the daughters of earth, things are the sons of heaven. How does this insight apply to nonliterate primitive hunters, preindustrial men and women—or museum curators who collect things?

He believed that the Coca-Cola bottle, discarded millions of times a day around the world, provides opportunities for carrying out many procedures used on "museum-level" objects such as prehistoric pottery, candlesticks, and domestic architecture. Don't all these items provide ways of informing the mind, and thus shaping the culture?

He got to work and proved his point. He charted the diagnostic attributes of Coke bottles, including family, types, and variants, and went on to study sequence and durations. The lowly Coke bottle can tell us a lot not only about public tastes, but the evolution of a commercial product.

The bottle, drink, and unsurpassed advertising-marketing strategy have combined to create a global Coca-Cola Culture. Coke is the American fuel, just as television is its soul. The world has responded with a resounding "Yes!," singing the Coke jingles and learning new ones as quickly as possible. We gave the world a Coke, and they drank it joyfully. Coca-Cola executives point out that their product has entered the lives of more people than any other icon or ideology, including the Christian religion.

Coca-Cola believes in the future of all nations, and will ride through whatever political or economic conditions exist. They are able, with this philosophy, to coexist with Hitler, impoverished migrant workers, Guatemalan death squads, hard-nosed dictators, or repressive Chinese.

What seems to bind the people of the world together today is not a flag, language, or ideology, but brand-name consumer products. An opulent imaginary world has changed our perception of reality. We have new gods. Call them icons of capitalism.

Coke bottles actually did get into museums—through the work of pop artists such as Robert Rauschenberg and Andy Warhol. The first "pop" depiction, Rauschenberg's "Coca-Cola Plan," was completed in 1958. Andy Warhol's "210 Coke Bottles" followed in 1962. Since then the bottle has appeared in innumerable pop paintings, collages, and advertisements.

In recent years a series of pop icons have surfaced, but none have equaled the power of the Coke bottle. Other questions remain to be explored. What psychological qualities contribute to the iconic effectiveness of the Coke bottle? Are there ritual behaviors that go with Coke Breaks? Can one compare the changing historic roles of the Christian cross, the Communist sickle, the American flag, and the Coke bottle?

Apparently we have only begun to scratch the surface of pop iconology. Scholarship and criticism must catch up with life as we live it, and with the people who create it. They will be far different in the Information Age than they were in earlier times.

The Greeks had their mythology and Mount Olympus, the Romans their Caesars and forums, the early Christians their hagiography and catacombs. Medieval Christians had their crusades and cathedrals. In early modern times European nations had their empires. After conquering the eastern wilderness, and winning their revolution, Americans had their western frontier. Now that frontier has moved into outer space. Who knows what new icons await us there?

Golden Bangladesh

In Bangladesh, three great rivers—the Ganges, Brahmaputra, and Meghna—converge to make one of the cradles of civilization. The land, with one of the highest annual rainfalls on earth, has been cultivated and loved for centuries beyond reckoning. The Pundras were here before recorded history; the Buddha came here as a pilgrim. Ashoka, great Mauryan emperor, built a stupa around 250 B.C. to commemorate a relic left here by the Buddha. The kings who ruled the lower Ganges for centuries, great patrons of art and learning, founded universities and supported sculptors and painters.

These were happy and prosperous years. Cotton fabric, known as muslin, became widely renowned, as did their delicate work in gold, silver, and conch shells. Poetry, music, and dance flourished. Word of all this reached the Western world through the work of Hsüan-tsang, the Chinese traveler, who was there between 638 and 645 A.D. Now it was my turn to make this pilgrimage to Golden Bangladesh: home of the mighty Bengal tiger and ancient memories.

Most Americans know nothing of all this. To them Bengal Tigers are one of the beefier professional football teams. The ad makers have urged them to "Put a Tiger in Your Tank," and to smile at Tony the Tiger on their cereal box. Thus do ancient history and popular culture meet at the breakfast table.

Few know much of the history and anguish of Bangladesh, the eighth most populated spot on earth, and one of the poorest nations. Islam entered the area in the twelfth century via Arab traders and missionaries, after which the Portuguese, French, and English came to this lush green land. In 1757 the bold Clive of Plassey, with 8,000 irregulars, overcame 50,000 troops of the Nawab of Bengal, and established a British control that lasted for two centuries. The ancient Hindu-Muslim rift widened, and in 1909 the British had separate electorates for the two forces, but there was no real peace.

When India became independent in 1947, terrible religious conflict broke out and massacres occurred. Two areas a thousand miles apart were set up for Muslims, and named East and West Pakistan. The arrangement was doomed to bloody failure.

West Pakistan forced a new colonialism on the East, and decreed Urdu as the national language. Civil war erupted. The well-equipped West Pakistanis swept down to loot, rape, and burn. Cities were seething masses of flames in which people were roasted alive. When women tried to strangle themselves with their long black hair, they were shaved bald and repeatedly raped.

University dorms were attacked with tanks and flamethrowers. Surviving students were forced to dig mass graves for slaughtered classmates—then shot and stuffed into the holes they had dug with their own hands. Over three million people were slaughtered. Ten million refugees fled into India. Finally India intervened and repulsed the Pakistani forces. A new battered nation—Bangladesh—came into being in 1971.

Years later, I go to "Golden Bangla" as the first American Fulbright professor to be admitted in a decade. I sit on my clean spacious veranda, look out over the scarred city, and wonder about the human predicament. Where is mercy, justice, reason? Just beyond my immaculately pruned and pleached garden two women sweepers listlessly pursue a single fallen leaf. A dark, younger woman squats atop a huge pile of bricks. With a heavy hammer she smashes them into small pieces against a large stone—hour after hour, day after day. Even on Friday, the Muslim's day of rest, and on holidays she works, accompanied by her three already old-looking children, midget-sized, torpid, playing on the bricks. Their eyes stare out of deep sockets; flies swarm around them. Her hammer comes down like the hand of fate. Finally the small pieces form a large mound. Then other skin-and-bone women come, put the pieces in baskets, and carry them off on their heads.

The Brick Woman has become my symbol for Bangladesh. Policies, programs, and projects change and she keeps breaking bricks. What does she think about, day after day, hour after hour? Where do she and her children go at night? Does she dream of a better life? What can she expect? How long will she live? I have no answers. Perhaps she has no questions.

I write about her in my clean white, air-conditioned room. Soon the bearer will come (in his clean white coat) with my afternoon tea (in a clean white cup). When I am tired I can nap (on clean white sheets). Outside, the Brick Woman keeps pounding.

* * *

A visitor to Bangladesh can never forget and never escape the rickshaws, called little taxis. Most of them—perhaps 95,000—are bicycle driven, and thousands more are gasoline powered. Rickshaws are not only a way of life, but a way of art—a traditional art made by poor workers for poor consumers, glorifying a brilliant nature and a tumultuous past. The first adequate account of this remarkable folk art finally appeared in 1997—Professor Henry Glassie's *Art and Life in Bangladesh*. He speaks, lives, and works with the artists.

If rickshaws had voices they would sing. Instead, they prance and parade—moving art galleries displayed in paintings, collages, cartoons, and cut-outs. They feature parrots, peacocks, swans, deer, jackals, crocodiles, snakes, tigers, lions, and elephants. The favorite flower is the lotus, an emblem of Bangladesh, shared by Buddhists, Hindus, and Muslims. The most displayed building is the famous Taj Mahal.

One looks at all this in amazement as the rickshaws move past. On one the lion, king of beasts, delivers a speech to a guitar-playing monkey. On another three-faced winged snakes are breathing fire. Animals wield guns, smoke cigarettes, and take command. There is a vivacity and creativity that takes one's breath away.

Untrained but undaunted, rickshaw artists let their imaginations fly free. If they run out of red paint, they finish with another color. Suddenly, a blue sky becomes half red, and a blue ocean becomes half yellow. Fat first world cows graze on sparse third world meadows. A mosque stands beside a skyscraper. Roads skim over lakes, trains fly alongside airplanes. Folk artists may live in humble surroundings, but their work dwells in realms of gold. Henry Glassie says it best:

> The soul is the breath of God, caged in the body, and art is its song. It is the inner force that unifies the human being with nature and with God.[1]

In Bangladesh, there is misery and poverty, but no proletariat. The true hallmark of the proletariat is neither poverty nor humble birth, but a consciousness of being shut out—of being disinherited, unwanted, in a community which is their rightful home. In places such as Bangladesh, where folklore, folk art, legends, rickshaw decorations, community, mythology, drama, and dancing abound, there is a common culture but no Marxian proletariat.

This does not mean there is not anguish, strife, and bloodshed. Islam entered the area to rule by the sword, and the never-ending conflict with Hinduism began. One tragic conflict follows the last. Here nature truly is red in tooth and claw. Anyone who wants to see life on the dirt level—life in the raw—should come to Bangladesh.

Charles Darwin spent five years on the *Beagle*, formulating his theory about the survival of the fittest. One might learn the same lesson in a Bangladesh garbage dump—overnight. First come starving people, to pick over and swallow anything they can. Then come the dogs who devour bones people can't handle. After the dogs come the cats; then the rats; finally the roaches and the ants. By morning, everything will be gone.

Here people don't live lives of quiet desperation—the desperation is noisy. Day and night, winter and summer, streets teem with love, hate, sobs, laughs, birthing, dying. Survival is the only game in town, and everyone plays it.

Large black crows fill the skies. They seem even more ominous than packs of dogs, which look like a cross between the ferret and the fox terrier—half teeth and half body, fast as rabbits, always ready to attack. Some are rabid, so no one dares bother them—not even the carrion crows.

The things I see and smell in Bangladesh are like much of the third world, where most people are born poor, live harsh, hungry lives, and die young, leaving poor children in their wake. Out of a hundred children born every minute in developing countries, twenty will die within the year. Of the eighty survivors, sixty will have no access to modern medical care. Of those who live to school age, only one in seven will ever set foot in a classroom.

What is it like in rural areas, where more than 90 percent of the people still live? I get on a battered bus and set out to see. There are few roads, and no highways, as we define them. The wet, narrow, poorly paved lanes, partly washed out by heavy rains, are suited only for bullocks and bicycles. Down them dash trucks, buses, ancient cars, overloaded pushcarts, migrants, merchants, bullocks, gurus, and goats. There are no speed limits, no guard rails, no police patrols. Only one law prevails: The Law of the Horn. Without traffic lights, road shoulders, rest stops, or turnoffs, the horn prevails. Traveling is like blind man's bluff, with human life at stake. Gabriel, blow your horn!

Luck is on our side. We reach Chittagong alive—a polyglot commercial port coming apart at the seams. A beautiful British World War II cemetery is here. Hundreds of servicemen who died in nearby Burma are buried in a green field overlooking the blue Bay of Bengal. The grounds are beautifully kept, and each grave is marked with a name, serial number, rank, and brief epitaph.

In the center is the statue of a large sword, shaped like a cross—waiting for some King Arthur to wield it once more and right the world's wrongdoing. An aging Indian caretaker, called Lame Bailey, greets me and manages a snappy British salute.

> Good men all they were, Sir. I fought with them, and I know. We were the proud Bengal Lancers. We put on a bloody good show. Every night I come here to guard and salute them. I will never leave them, my comrades. They are a bit lonesome. A long way from home and family, you know, Sir.

There are tears in Bailey's tired eyes—and in mine.

* * *

In a pensive mood, I leave the cemetery and go to a nearby spot on the Bay of Bengal called Cox's Bazar—the longest stretch of unspoiled beach in Asia, mile after mile of clean white sand lapped by clear, blue water. In the distance a lone figure sits facing the water. I draw closer. A wiry bent man, wearing the Buddhist monk's saffron robe, sits cross-legged, staring at the glassy sea. He beckons me over.

"Come sit with me and empty yourself, as the rivers do. Here there is sunshine and peace."

I sit. He smiles and speaks. "You are a stranger and are welcome. This land is good to strangers."

"You are a Buddhist monk?"

"Yes. Captain Cox, a British naval officer, brought Buddhist refugees here in 1798. He liked it so well that he stayed to set up Cox's Bazar. We stayed too, and formed a thriving Buddhist colony and monastery. Now this is our home. Golden Bangladesh."

"Do you get lonesome here on the beach?"

"I am alone, but I am never lonesome. I greatly enjoy sitting here, letting myself be baked in the sun till I am done. When I have ripened, I shall die and be born again."

Part of me welcomed what the old monk said. I wanted to sit quietly with him and ripen in the sun. At the same time, something inside me rebelled. The West inside me rejected his East, with its resignation and quietude. Lines from a school poem pop up: "Let us then be up and doing, Life's a thing will not endure."

Our bus driver will be honking his horn. I hurry back to Chittagong, prepared to drive north to Dhaka at breakneck speed. For five bone-jarring, careening hours, I recall the silent serene beach. Had I left too soon? Did I pass up my only chance to turn my life around? Will I never get rid of what Buddhists call "the mad monkey of the mind"? Will I ever ripen?

Horn blaring, we pull into Dhaka's crowded noisy depot. How many times have Dhaka's soggy streets been orange with fire, black with ashes, red with Bengal blood? How have these people not only survived, but held on to their ancient popular culture?

On my last night in Bangladesh, I decide to take a long walk. Perhaps I will see the Brick Woman, cart pushers, singing schoolchildren, or rickshaws with garish, glorious images. The land is full of surprises, and every walk is an adventure.

Heavy rains come. Empty rickshaws cruise slowly by, their teenage drivers dripping like ducks in ponds that have no dry banks. They glide silently by with little hope of customers.

Suddenly the silence is broken by a pure piercing tenor voice, singing as if all the world were his stage and humanity his audience.

A Bengali folk epic fills all space, passed on orally for many generations, now meant just for me in 1998 A.D.

The frail singer emerges from an alley that enters my street at a right angle; a blind troubadour, like those who have no memorial, yet gave life meaning through singing long-cherished songs from their disappearing past.

How or why is he here? Who can say? He is a thing of rags and tatters. His bare, bruised feet have hit many a sharp stone, causing daily, even hourly, pain. Having neither guide, friend, nor cane, he runs into obstacles in his path. Yet he never slows down.

Passing rickshaws splash mud and cow dung on his frayed lungi. A victim of smallpox, his body is deeply pitted with ugly scars on his eyes, cheeks, neck, and even his closed eyelids. Sores on both legs tell how many battles he has lost in his eternal darkness. One on his right thigh has not healed: a festering open sore. Every rib shows through his almost-fleshless torso. Still he fills his small lungs and sends forth sweet sounds. He has found his glory.

His rain-soaked head held high and triumphant, he moves toward me, ever more exultant. What a beautiful man—he is singing of heroes, lovers, monsters, chariots of fire. In his cupped right hand are small coins others have put there—paisas worth less than one taka—four cents in American money. The coins from my pocket almost fill his hand. His already smiling face becomes radiant. The song goes on, uninterrupted. He walks on to his predestined fate, but his song stays with me, as if it were my own. How could this be? Had I known it in an earlier life, or heard it in a childhood dream? If the child is father of the man, am I being blessed by what Wordsworth called natural piety? When might I hear the wonderful music again? I shall have to wait and see. As for now, I have this moment to cherish.

I am keenly aware of all the good things we have, and Bangladesh needs. Of course I would like the sick, blind, and lame to enjoy modern medical treatment. Perhaps Lame Bailey could have walked, the Blind Troubadour could have seen and his smallpox been cured or alleviated. Social conditions cry out for change—my Brick Woman and her children must not face endless exploitation. With adequate birth control practices, she could control the size of her starving family. Filthy water can be made safe. All phases of the educational system can and should be updated. Women must be treated as humans, not slaves. We could go on and on. To allow the current trend at home and abroad—for the rich to get richer and the poor poorer— will guarantee not progress but revolution.

Nothing comes without a price. Change can itself invite chaos and revolution—recent events among Bangladesh's neighbors in southeast Asia prove the point. In imprinting the world with *our* version of progress, we must heed the warning of Walt Whitman to be bold—but not too damn bold.

It's easy for people in the so-called "developed world," adrift in their electronic utopia, to forget that for most people of the world, our new self-styled Electronic Revolution has little or no meaning. It negates customs, rituals, and religions that are deeply imbedded. Many have no burning desire and no resources to enter our version of the New Utopia. With an average yearly income far less than the price of the cheapest computer, their desperate need is not to be on-line, but to stay alive.

The economic gap between the rich and the poor involves much more than income. The technological gap is even more conspicuous—as some nations advance, others may be left even further behind. Nations in Africa, Asia, and Latin America found out in 1998 that instead of entering the global village, they may be increasing the wealth of their slumlords and magnifying their global slums.

* * *

While one is sitting in a comfortable American office with a high-speed computer, the idea of showering Bangladesh with our new technologies might seem to be an obvious blessing. Forget the fact that it backfired in Thailand, Malaysia, and Indonesia. Those who actually live and die there know that such societies have exceedingly complex social, ritual, and artistic ways of life. Why do we think they are willing to surrender them?

A notable feature of Bengali society is that full peasant participation functions practically without urban centers or electronic gadgetry—markets, trade, varied services, social and political hierarchy, and religious hierarchy can all function without cities. When the British left Bangladesh it was barely 3 percent urban; by 1961 it was only 5 percent, and today it is less than 12 percent—one of the lowest rates in the world despite the population density. While neighboring India and Pakistan develop atomic bombs, the Bengali kinship pattern remains atomistic, overlapping, and intertwining, with multiple meanings few Westerners have ever tried to understand.

We have difficult questions to answer. What happens when electronic mass culture invades and swamps a traditional oral-aural culture? What do we need to know to make an intelligent and humane answer? If we tread softly, must we carry both a big stick and a big computer?

Dr. Syed Anwar Husain, widely regarded as Bangladesh's leading historian, is not sympathetic with what he calls the hard-headed policymakers in the United States. He notes that the traumatic effects in the 1960s of our Vietnam fiasco allowed us to equip West Pakistan to crush East Pakistan, believing that the time span for the emergence of Bangladesh was at least five and up to ten years.

How, he asks, could a superpower such as the United States have such a paternalistic feeling for the Bengali nationalists? How could our estimate be so wrong, our policies so shortsighted?

In 1998, we debate our policies toward a newly emerging superpower, China. President Clinton assures us that we should overlook all but our business interests, and even ignores China's supplying nuclear information to Pakistan. We hear instead of how many American goods China buys, how much Coca-Cola China drinks. Let us hope we do not succumb to the glib get-rich-quick philosophy.

Am I being soft-headed and Rousseauistic, sitting in a posh Foundation House where "our" people come visit to "see" Bangladesh? The walls are white, the water is boiled, the food is Western, and everyone speaks English. My room has a pleasant veranda overlooking a grass lawn that would grace a first-class golf green anywhere in the world. Why then do I feel a deep uneasiness?

On the wall is a glossy picture of a Bengal tiger. He is not tarnished like his native land. Instead, he seems right out of the pages of William Blake:

> Tyger! Tyger! burning bright
> In the forests of the night,
> What immortal hand or eye
> Could frame thy fearful symmetry?"

"The Tyger," st. 1

* * *

ENDNOTE

By far the best book in English on Bangladesh is Henry Glassie's *Art and Life in Bangladesh* (Bloomington, Indiana University Press, 1997). In addition to splendid text and photographs, it has a fine glossary and index. See also my book *Bangladesh: Inter-Cultural Studies* (Dhaka, Bangladesh, Ananda Mudrayan Ltd., 1983); American University, *Area Handbook for Bangladesh* (Washington, DC, American University Press, 1975); Omar Chowdhury, *Prelude to Bangladesh* (New Delhi, South Asian Publishers, 1980); B. L. C. Johnson, *Bangladesh* (New York, Barnes and Noble, 1975); and Syed Anwar Husain, *Bangladesher Swadhinata Juddeh Parashaktir Bhumika* (Dhaka, Bangladesh, Dana Prokashoni, 1982).

Global Village

There is nothing new about global thinking and model building. Long before we knew the world was round, shaped like a globe, the idea of human unity—in a village, city, region, empire—prevailed. That idea, which has always affected art, religion, literature, and politics, is still a dominant dream of the late twentieth century.

People everywhere have prayed "that we may be all of one heart and of one soul, with one mind." That prayer has never been answered. But it has never disappeared either. Countless multitudes have dreamed and died for a utopian unity—a dream that changes but always eludes us.

What is left is ruined cities, monuments, and dreams. The poet Shelley described "Two vast and trunkless legs of stone," standing in an endless desert with this inscription:

> My name is Ozymandias, king of kings;
> Look on my works, ye Mighty, and despair!
>
> "Ozymandias"

Other mighty ones rise and fall. In 334 B.C., Alexander the Great swept into Asia. In eleven years he conquered most of the civilized world. In those same years a youth named Zeno founded Stoicism, holding that all men are common citizens in the same world state. While Caesar was repeating Alexander's conquests with his Roman legions centuries later, Cicero was writing: "We are all bound together by a common chain." That same idea took on a different name some two thousand years later when Marshall McLuhan popularized the phrase "global village."

Another force fostering the "one-world doctrine" has been religion. Roman Christianity, proclaimed by Constantine in the fourth century A.D. and later formalized in the papacy, dominated Western civilization before the emergence of nationalism. Other religions have continued to exert their strong claims—such as Judaism with its notion of the chosen people of God, and Islam, following Mohammed's exhortation to conquer by the sword. But religious wars, still raging in our own day, refute any hope of a universal religion.

For centuries the Holy Roman Empire (which Voltaire said was neither holy nor Roman nor an empire) gripped much of Europe. Global consciousness flourished in the Age of Imperialism, when European nations carved up much of the non-Western world. Large segments of the globe were ruled by the British in the nineteenth century, when "the sun never set on British soil." In recent times, imperialists' globalism has been countered by independence movements in India, Africa, the Caribbean, and the Middle East.

The same fate awaited Communism and Nazism in the twentieth century. The Third Reich, which Hitler boasted would last a thousand years, lasted twelve. Statues of Lenin and Stalin were toppled throughout communist Eastern Europe. Kings and crowns come and go. "Let us sit upon the ground," Shakespeare wrote, "and tell sad tales of the death of kings." Perhaps later historians will tell sad tales about the global village.

Its special feature is dependence on the Electronic Revolution and rapidly advancing technology. Sparta and Rome were military powers, Israel and Islam religious powers, Britain a naval power that "ruled the waves." Earlier on, armies and navies, Bibles and bombs, were tools for imperialists and empire builders. Today the real power dwells with the media—no longer on the waves, but the airwaves. The dynamo driving us forward is consumer capitalism. Not big guns but Big Mac seems to conquer friends and foes alike.

Obviously the world is shrinking: cheap air travel, tourism, sports, fast food, fast cars, film, television, theater, and popular spectacles ("world competition" for beauty, brawn, and glitz) flourish. Cloning and copyright theft abound. Tapes of American rock bands and movies are cheaper in Beijing than in Boston or Baltimore. We become increasingly similar (clothing, hit songs, traffic signs, class B Hollywood movies, for example, know no boundaries).

The story doesn't end there. External intrusion into cultures creates new tensions between the forces for the status quo and the new monolithic globalism. The result is a worldwide tug-of-war.

The more invasive the technology, the stronger the backlash. Ours is an age of global paradox.

Relations between nations and publics have become more complex, fragile, and sometimes hostile. Small but vicious wars crop up everywhere, causing havoc in Africa, the Balkans, and Ireland. Writing of these things in his book *Global Paradox* (1995), John Naisbitt predicts that governments will continue to fall apart (as did the world-girdling U.S.S.R.), moving to smaller units that can adapt quickly to change and preserve their indigenous cultures. On the home front, the battle will be with words rather than guns—we call them "culture wars."

Major questions emerge at the end of our tumultuous century. Will we choose convergence or divergence? Can we somehow have both? Is personal space public or private? Should we dominate or harmonize with nature? Is our goal to achieve or to be?

* * *

A new gold rush is on. All that glitters isn't gold—but glittering gold entrances and enslaves. Many Europeans came to and settled in the New World to get gold. The prize today is not in the metal taken from the ground, but in silicon chips, taking information from the air. Today's "richest man in the world," who can and does build a $50 million home, is Bill Gates. What is the name of his gold mine? Microsoft.

When did the Electronic Revolution begin, and what role is it playing in the world poised on the brink of the twenty-first century? Benjamin Franklin was performing his electrical experiments before the American Revolution; but the dramatic breakthrough came with the "three Ts" (telegraph, telephone, typewriter) in the nineteenth century. The telegraph gave us a social nervous system, divorced from "muscle culture." By making information available instantaneously, it effectively ended the separation of transportation and communication. Telegraph messages were sent around Europe in the 1850s and across the Atlantic in 1866. The telephone and typewriter came soon afterward, transforming the culture. Edison's invention of the electric lightbulb soon lit up the world. The "Wizard of Menlo Park" is the godfather of today's Electronic Revolution.

The advent of radio made possible simultaneous communication between a central authority and any number of receivers. The union of invention and progress in scientific and technical areas provided a cornucopia of new methods, devices, and systems. But nothing was so startling and revolutionary as the computerization of the late twentieth century.

The first successful computer, nicknamed ENIAC, was built at the University of Pennsylvania in 1946. In 1950 there were fewer than 900 computers in America—and only 2,000 ten years later. Then the numbers soared—to over 125,000 in 1970 and half a million by 1985. In another decade, there were several million, at home and abroad.

With new computers came new prosperity, instant millionaires, and overwhelming hype. Bill Gates' face leered at us from adoring magazines, full of slick ads and utopian promises. The Clinton administration led the pack; Al Gore became a "computer in every classroom" drone. Only a few asked, if the unexamined life is not worth living, is unexamined technology not worth having?

Might we be victims of a glitzy fad? Fads and fancies dance to their own rhythms. They can no more be explained than sudden changes in the weather or clothing styles. The more we study popular culture the more baffling they all become.

Might a global village create global slums? What about the billions who would not use or share the new technology? Will the new sweatshops be in the world's Silicon Valleys? Will all nations accept our virtual reality and canned entertainment?

Anne Cooper-Chen's *Games in the Global Village: A 50-Nation Study of Entertainment Television* (1994) suggests an answer. No game format has a universal audience. Many "popular talk shows" appall other nations. Cultural preference follows patterns that the Electronic Revolution does not change. This in turn raises another crucial question: how can deep cultural variation be squared with the biological unity of the human species?

Earlier glowing accounts of the global village have been too simplistic and superficial. They underestimate the power of history, tradition, language, and autonomy. Billions of people still live as did their ancestors for untold generations: in tribal, local, regional, or national patterns. These tight communities cling to their limited horizons, despite intrusions from the media on the ground and satellites in the sky. Most people will travel no more than a few hundred miles—if that far—and will be buried close to the spot where they were born. Myths manipulate the world more than media. They always have.

Not only does globalism inherit the old myths; it also invents new ones. That is a major function of computers, television, films, magazines, advertising. How all this will mix with the old mythology no one can yet say. Will it increase or decrease our hostilities? Will it bring us together or tear us apart? The agonizing situations in places such as Bosnia, Iraq, Ireland, and a number of African countries does not bode well.

In Europe, which dominated global affairs for centuries, these are trying times. John Newhouse's account, called *Europe Adrift* (1997), suggests that the only thing unifying Europe today is unemployment. Instead of curing it, globalism exacerbates it. In Germany more people are unemployed than in the early days of the Third Reich; in France, not only blue-collar workers but also large numbers of white-collar workers join the ranks of the unemployed. The "safety nets" of jobs and security disappear in country after country. Europe's real strength rests not in nations but in two regions, both shaped liked bananas. The first stretches from southern England through Benelux into Switzerland. The second begins in Italy and follows the Mediterranean coast westward to Barcelona. The idea of either a Europeanized Germany or a Germanized Europe has faded. If Europe cannot effectively unite, how can it hope to lead a united world?

Nor does the firestorm raging through Asia augur well—it will leave no country untouched, predicted the International Monetary Fund in a report issued at the end of 1997. Economic growth will slow and unemployment will rise, especially in nations at the center of the crisis.

Certainly people everywhere will be bombarded by more sights and sounds than were their ancestors. Much of this will be in languages they do not understand, about places they never expect to see. What effect will news of the world have on them? Colin Cherry, professor of telecommunication at the University of London, thinks it may not unite them since the spread of world news may do as much to keep people apart in their emotion and attitudes as does to lead to peace and understanding.

To put it differently: we have a booming global economy but we don't have a global society. Markets reduce everything (including human beings and nature) to commodities. Societies need more than this to prosper—such goals as political freedom and social justice.

In the age of mergers and buyouts, global corporations and global advertising agencies are changing the way we receive news and entertainment. A handful of mammoth private organization are dominating the world's mass media. Five or six corporate giants control most of the world's major TV and radio stations, newspapers, magazines, books, recordings, and videocassettes. Ben H. Bagdikian, a leading student of these matters, warns this does not bode well for McLuhan's "universal understanding":

> The lords of the global village have their own political agenda. All resist changes that do not support their own financial interests. Together, they exert a homogenizing power over ideas, culture and commerce that affects populations larger than any in history.

Still there are great limitations to this power, much of which depends on corporate buyouts and mergers. Corporations come and go—nations and national interests endure, despite changes in names and boundaries. The central cord that ties people together is not the media, but the belief in our ability to choose our ends for ourselves. That is the real bottom line.

The Western world must not assume that its cultural patterns will suit everyone. Consider the difference between our democratic system and that of the world's largest democracy, India. For centuries India has been and (despite some modifications) remains a stratified society with a rigid caste system. When a worldwide beauty pageant with scantily clothed women was held there in 1997 the protest was violent, and some Indian women actually burned themselves to death as part of that protest.

In China, the world's most populous nation, it is all too clear that our ideas of freedom and democracy will not work in a Communist government where the media and the masses are strictly controlled.

Our relations with other Asian nations have become more strained, and economic patterns ruptured, in places where loyalty, not conflict, is honored; and where ethical practices not acceptable to us are deeply entrenched. In many nations, especially in Latin America, bribery and nepotism have a long history. So do class, gender, and clan practices quite unlike ours. Islamic women, for example, live by very different rules from those of most American women. So do many others around the world. No flashing messages on the Internet will change this overnight—or for years to come. Our roots go very deep.

True enough, technology and capitalism are sweeping the globe; but capitalism is not only an economy but a culture, with many distinctive variants. A book by C. Hampden-Turner and A. Trompenaars identifies *The Seven Cultures of Capitalism* (1995). My own travels and observations lead me to believe the number is more like seventy times seven. And with the new stress on diversity, the number is growing.

We can transfer money and stocks in the wink of an eye—a process that has wrought financial chaos in Asia, and sent the world markets reeling. The continued importance of national currencies, as the European Union has discovered, undercuts dreams of globalization. The impact of the Euro is uncertain. The global village is a utopian vision; the reality is the nation, region, and locality, which still control and sustain us. Millions of messages and scores of satellites circle the globe. Yet we yearn to give to airy nothings a local habitation and a name.

Where have we gone astray? What happened to the euphoria of recent years? History happened, as it always does. We let our pride obscure our prejudice, substituted hubris for human limitations, falling into the trap our greatest thinkers have always warned us we must avoid—Sophocles, Aeschylus, Augustine, Dante, Shakespeare, Milton, Melville, Emerson, Faulkner, and many others. We have allowed the rich to get richer, the poor poorer. We have confused glamorous record profits with record achievement. The consumer capitalism with its relentless and heartless strategies will eventually consume us too. Like every other boom, there will be the inevitable bust.

Is there life after the fading of global village optimism? Absolutely—we can count on it. The pendulum will swing and we will have new programs. Life in the twentieth century has been a roller-coaster ride: ups and downs to take away our breath, our balance, even our lives. The Baby Boomers, with their own kind ensconced in the White House, will have their day, and then their children, the Baby Busters, will have theirs. Theirs will be no easy job. Born of parents who had the highest divorce rate in history, saddled with debts in the trillions, massive pollution, disintegrating families and cities, racial rancor, they will find little solace in the disappearing utopia. Twenty-nine-year-old Robert Lukefahr summed it up, saying they feel like a generation of janitors.

Perhaps we tackled too much too quickly—the great American impatience. First of all, following President Clinton's lead, we must put our own house (budget, medical system, bureaucracy) in order. Instead of standing aside and attacking the system, we might work for change from within, demanding both public and private-sector support. We have seen the enemy and he is us.

As for the global village, perhaps it was—to use the title of a movie about the Allies' unsuccessful effort to rush prematurely into Germany—a bridge too far. Instead of thinking global, I suggest that we think regional.

A region may be defined as an area held together by a common culture—myth, language, religion, a sense of belonging. The key to a workable region is cooperation, not (as with sections or nations) hostility and antagonism. Instead of attempting to monitor and regulate the scores of nations now in the UN, we might work through regional councils and parliaments, which would take some of the terrible pressures off the almost-bankrupt and overtaxed UN.

Our own United States provides a model. We have three overarching regions the North, South, and West, and people are proud (sometimes too proud) to acknowledge this. But there are distinct and workable subregions within the three—New England and the Middle Atlantic in the North; Upper and Lower South, Midwest and Far West. More recently we have spoken of "belts"—Sun Belt, Rust Belt, Corn Belt, and so on. In this sense we can say we have not one but several "Americas," regions in search of a common faith and destiny.

How might this work if we put the grid on the whole world? The overarching "regions" become obvious: Western Europe, Eastern Europe, North America, Latin America, Pacific Rim, Southeast Asia, Middle East, South Pacific, Northern Africa, Southern Africa.

Many would want to fine-tune the list, or use different labels. But the implications are clear. Western Europe has more than enough to do to make the EC work; North America, NAFTA; the Middle East, the Israel-Palestine accord. All problems eventually become global; but by attacking them first on a regional level, and avoiding utopian overexpansion, we might get much better results.

What is needed is not to restore the "good old days" of national rivalry and cutthroat competition, or to revive a vanishing culture under modern conditions that make it impossible, but to grow a contemporary world culture from the old roots. The crucial problem is that each region preserve its characteristic culture, which should also harmonize with that of its neighbors. We cannot give up our heritages with al their rich variety. A diversification of human communities is the foundation of civilization.

ENDNOTE

Books about the global village and its new technology form a library of their own. Were I to choose half a dozen books that will have permanent value, they would be these:

1. Paul Mantoux, *The Industrial Revolution in the Eighteenth Century* (New York, Harper Torchbooks, 1961).
2. Lewis Mumford, *Technics and Civilization* (New York, Harbinger, 1963).
3. Jacques Ellul, *The Technological Society* (New York, Knopf, 1964).
4. Theodore Roszak, *The Cult of Information* (Berkeley, CA, University of California Press, 1994).
5. Neil Postman, *Technopoly* (New York, Knopf, 1992).
6. Scott Adams, *The Dilbert Principle* (New York, HarperCollins, 1996).

I know of only one book that has attempted to establish a new field of study—comparative popular culture. It is Roger Rollin's *The Americanization of the Global Village: Essays in Comparative Culture* (Bowling Green, OH, Popular Press, 1990). This new effort might supply the missing link to our earlier studies.

PART IV: LINKS

East and West

In 1974 I flew over the wide Pacific to the flourishing East-West Center in Hawaii to lecture on popular culture and the New Order. My message emerged loud and clear: The global village is the new reality. Boston is linked to Beijing, Pretoria to Peoria, New York to New Delhi. Welcome to a flourishing global village.

For sheer inventiveness, I pointed out, electronics have no equal in history. A person may well have lived from the advent of the radio to the landing on the moon. He or she could have seen the birth of atomic power, jet travel, television, Sputnik, and the Space Age. The computer might alter our lives as much as the printing press did in earlier centuries. Bill Gates might become the new Gutenberg. Who can say?

Unprecedented scientific and technical progress provided a cornucopia of new methods, devices, systems; old ideas and new were turned upside down; many models were obsolete before they were put into place. O brave new world!

The scholar who best understood what was happening—he invented the term global village—was the new Prince of Pop, Marshall McLuhan. Everyone wanted him to speak, write, predict. I got the message from McLuhan himself, and was a true disciple. McLuhan buzzwords filled my talks—experimental, interface, dynamic, nebular, nuclear, freefall, random, hot, cool, global. East and West were wedded. We had outgrown Rudyard Kipling's famous poem, "The Ballad of East and West":

> Oh, East is East, and West is West,
> And never the twain shall meet . . .[2]

We had amended and expanded the most famous doctrine in American history—the Turner thesis. Three short essays by Frederick Jackson Turner, published at the close of the nineteenth century, changed the direction of American thought.

Turner argued that the frontier was not so much a place as a process, creating a new person, the American. The grand sweep of settlements westward meant a steady retreat from Old World influences and a steady growth of independence. We went west first as hunters, then traders, ranchers, farmers; finally we produced the city and the factory. The frontier was the line of most rapid and effective Americanization.

We entered the Appalachian frontier with coonskin cap and long rifle, to emerge on the Great Plains with ten-gallon hats and six guns—our Manifest Destiny. It was the stuff that myths are made of.

Yet even as Turner was writing the frontier was closing. The 1890 census showed that western settlements "were so scattered over the region that they could no longer be called a frontier." The first period of American history was closing. Now Americans would have to find new frontiers.

The frontier thesis was relatively simple, easy to understand, affirming national pride and purpose. Democracy, it held, grew from a series of adaptations to new environments. With an abundance of free land and open space, there was a "safety valve" that gave an outlet for dissent, and ultimately to America's political system. In the New World people were freed from the danger of a landed, thus perpetuated, aristocracy.

Turner died in 1932, but his thesis, like Old Man River, just kept rolling along. By then the frontier had become mythic, encapsulating the opposition of good versus bad, East versus West, nostalgia versus change. The thesis had been applied not only to American history but (as in the work of scholars such as M. W. Mikesell, M. T. Katzman, W. T. Jackson, and A. Hennessey) to other frontiers all over the world. In a larger context, the whole New World was seen as the frontier for the Old World for nearly four centuries. As one frontier closes, another opens. Thus newly elected President John F. Kennedy proclaimed the New Frontier in 1960.

Some say the frontier has since then moved to outer space; but I suggested we forget the old frontier, that we seek it elsewhere—to the far west of America's Far West—across the vast Pacific Ocean to what we have long called the Far East.

The Pacific Rim is our new frontier. This was my explanation.

This Rim (or Basin, or Community) embraces the Asian Pacific, coinciding roughly with the western rim of the Pacific Ocean. Because of their unique importance, Japan and China spring first to mind. One then thinks quickly of the newly industrialized countries (NICs): South Korea, Taiwan, Singapore; of the Association for Southeast Asian Nations (ASEAN): Indonesia, the Philippines, Thailand, Malaysia, Brunei, and (again) Singapore; the far-flung islands of Oceania; and of the land controlled (in 1974) by the U.S.S.R. or socialist partners (Vietnam, North Korea, and Cambodia). Australia and New Zealand are somewhat removed from Asia by their location and Western orientation.

The growth and changes in this part of the world since World War II have been dramatic and revolutionary. Pacific Rim economic growth has far outdistanced that of the rest of the world, leading to gigantic trade deficits and talk of protectionism. Without warning, the "miracle" of the Rim ended. Asian economies collapsed, or went bankrupt. Suddenly all eyes turned East, where the drama of our expanding world is writ large.

In the 1970s, more and more representative American groups, in business, art, education, and communications, had loosened their link to Mother Country. "Globalization is the answer to all our problems," I concluded. "This is the main road to industrial progress and peace. Let's all go down that yellow brick road together!"

Polite applause, the usual chit-chat, and we all went out to enjoy a hearty Hawaiian luau.

* * *

A quarter of a century later, I look back on that visit, and my talk, in disbelief. How could I have been so naïve? How could I have ignored the real problem Asia faced—the Vietnam War?

Many have told and retold the agony of that war. Year after year the fighting and the protests continued. Finally we had to withdraw. The bitter taste the war left not only in the United States but around the world altered our whole posture. Even after our military withdrawal, the fighting in neighboring Cambodia dragged on, leading eventually to genocide under Pol Pot and more years of dissension and bloodshed.

The "economic miracle of East Asia" could not hide the increasing troubles and ferment in many other parts of Asia. Further south, in Oceania, there were ethnic tensions in New Caledonia. The new government elected in Fiji in April 1987, was overthrown in May. Papuan separatism plagued West Irian. Insurgents opposed Indonesian rule in East Timor. Further north, in the Philippines, the victory of Corazon Aquino had not quelled the revolt of the New People's Army and Muslims, and the future seemed uncertain. The long-term question of reuniting Taiwan with mainland China was still unresolved. Highly visible and recurring trade tensions between the United States and Japan heightened, and spread to other Asia-Pacific countries. The 1986 Jenkins bill (vetoed by President Reagan) would have restricted textile imports from virtually every Asian country; that idea still had strong congressional support. President Bush's announcement, in April 1989, that America's new FSX fighter jet would be built in Japan raised a storm of protest. The debate over the reunification of the Korean peninsula, and America's military presence in South Korea, continued. Students were crushed in a shocking reaction to demonstrations in China. Everywhere the growth of ethnic and religious identity introduced explosive new issues and pitfalls.

I shall not attempt here to analyze these complex and far-ranging matters. Instead, I shall ask if, how, and when we shall find common ground and understanding on a global scale; if East and West can hope to meet; and if popular culture, subject of my special concern, can help.

Social, political, and cultural affairs did not seem to be fulfilling global village expectations—but at least economies were prospering—until the late 1990s. Country after country sent out distress signals. The economies of Thailand, Malaysia, and Korea were going bottom up. The worst news came from Indonesia, where a corrupt regime reluctant to make the necessary painful economic reforms suddenly faced large-scale riots, lootings, and killings. Hundreds of people were burned to death in a Jakarta mall that was being looted. Only martial law restored an uneasy truce.

In June 1998 came the most stunning news of all. Japan, the powerhouse behind the whole Asian revival and the second largest economy in the world, was officially in recession.

Things were getting worse in the Middle East and in the area formerly called the U.S.S.R. Russia, stripped of its many satellite states and communist governmental controls, was moving from bad to worse. Only a large amount of emergency funding from the World Bank kept it solvent. Humpty Dumpty had a great fall.

No one seemed to know what to do about the Balkans. Ethnic cleansing resumed, and incredible brutality went unchecked. The United Nations continued to rattle the saber, but refused to draw the sword.

More startling news followed: both India and Pakistan conducted nuclear tests, and demonstrated that they were prepared to join the "nuclear club" formerly limited to the United States, Britain, France, Russia, and China. The world seemed on the brink of altering its objective. The goal would be not peace but the sword—or the bomb. In the United States, a divided Congress seemed more willing to squabble than to legislate. China refused to modify its stand on human rights, and to relinquish its communist principles. It had supplied technical assistance to Pakistan in constructing nuclear weapons—and no one knew what to do about it. The East and West had not joined hands; instead they seemed to be at each other's throats.

No one knew where the world was heading, and whether or not its only superpower—the United States—could put Humpty Dumpty together again, closing alarming gaps both at home and abroad.

* * *

The most alarming gap is between the have-mores and the have-lesses. As our American production has gone up, workers' real wages have declined. The income of the top 1 percent of the population, who control 20 percent of the wealth, rose 78 percent between 1977 and 1998. Ours is a society of diverse economic groups—suspicious of both the future and each other.

A 1995 visit of Pope John Paul to the United States dramatized the economic disparities in our country at a time when the O. J. Simpson frenzy was consuming millions of dollars. What of the poor, the jobless, the homeless, the Pope asked? How long can this inequality continue? Was he implying that Humpty Dumpty might be headed for a fall? Did we not know that in 1993 our social health index hit an all-time low in six categories: children in poverty, child abuse, average weekly earnings, health insurance coverage, health costs, and the gap between rich and poor?

The time for reconstruction and renewal has come. We must lift up our hearts, reviving the great tradition that sustained us for centuries. We must master our new technology, lest it master us. We must learn to separate information from knowledge, and knowledge from wisdom. With lots of pluck and some luck we might be able to rescue Humpty Dumpty before he falls. We may not get a second chance.

* * *

ENDNOTE

Can the East and West ever meet? For centuries people believed, with Rudyard Kipling, they could not. Racial, cultural, historical, and religious traditions would see to that. A classic statement of the problem was F. S. C. Northrop's *The Meeting of East and West* (New York, Colliers, 1946).

The world shrank, technology conquered space, and when Marshall McLuhan's idea of a global village surfaced, many believed the twain would meet, creating a new world utopia. See McLuhan's *Understanding Media: The Extensions of Man* (New York, McGraw Hill, 1964). A strong attack on the McLuhan school appeared with James Grunig's *Decline of the Global Village: How Specialization Is Changing the Mass Media* (New York, General Hall, 1976). The remarkable growth of Far Eastern countries came to an abrupt halt in 1998, when utopian dreams began to disappear. For continuing articles on this topic, see various issues of the *Journal of Popular Culture*.

Everyman to Everyfan

Religion is the servant of humanity. Everyman has always wondered where he came from, where he is going, who is in charge. He turns to religion (from the Latin *religare*, to tie back, to rely) for possible answers. He performs the service and worship of God (or gods) and the supernatural.

God has always been popular in America. He sent out many of the settlers in the first place—or so they firmly believed. Americanism has, in itself, become a kind of popular religion. We are often said to be one of the most religious countries in the world—whatever that means. The stern mandate remains: We have invented a number of new religions, one of which, Mormonism, is among the fastest growing in the world. Our mandate has been clear: Get Right with God.

The mandate is, or course, worldwide. Few people have taken their gods so seriously, and treated them so reverently, as prehistoric people. By the process of apotheosis, many cultures (such as the Egyptian, Greek, and Roman) have turned men and women into gods and goddesses. Religion has been omnipresent and all-powerful.

How have we survived all these millions of years, and ended up in the Age of Information? Biologists point to Darwin, and the survival of the fittest. Theologians point to heaven, and say it is the will of God.

All societies have religion, and religion has always been popular. Society seems to be a religious phenomenon. The substance of culture is religion and the form of religion is culture.

Hold this thesis up to the twentieth century. Many believe the City of God has become the Secular City. Instead of depending on the Almighty, we depend on technology to solve problems. Forget ultimate goals; go for the practical and expedient.

Yet millions will not abandon the City of God and the Heavenly City. While "liberal religions" decline, fundamentalism and the "religious right" flourish throughout the world. Christianity, Judaism, Islam, and Buddhism attest to this. Powerful movements are affecting not only religion, but politics and public opinion. In Israel, India, Tibet, and the United States—to name four obvious examples—power centers must deal with fundamentalism to survive. In America the Moral Majority reversed the sixty-year hold of New Deal Democratic politics and ushered in a new era.

Tracing the tide of fundamentalism on the world scene is a vast task still to be accomplished. For our mosaic, I shall attempt to deal only with my own country, summarizing my longer account, *Great Awakenings: Popular Religion and Popular Culture* (1995).

Twentieth-century "ordinary Americans" struggle with questions that bothered our seventeenth-century pilgrims and dissenters. What happens to sinners in the hands of an angry God?

Our great awakenings occur about every fifty years. The first came in the middle of the eighteenth century, led by George Whitefield and Jonathan Edwards. It struck deep chords in the American psyche, which are reemerging here and around the world. Modernism is challenged by fundamentalism. Can one who isn't "born again" anticipate where this movement might lead? We wonder.

Look back now on our first great awakening, and ponder how the next one might alter American life in the twenty-first century.

* * *

Most Americans, like the earliest English settlers on our shores, consider themselves the chosen people, building a new haven on earth. Just as Moses guided Israel, so did Washington and Lincoln lead us—"By you all the families of the earth shall bless themselves." Ending the century as the world's only superpower, as well as military, scientific, and economic pacesetter, we have developed the Captain America Complex. Super Power, Super Dome, Super Bowl, Super Fly—elements of the new American religion.

Observe it on bumper stickers, speeding along our super highways: Smile, God loves you. America the Mighty. Love it or leave it. Honk if you love Jesus. We're number one!

And yell if your favorite team (baseball, football, basketball, hockey, tennis, lacrosse—male or female, big league or little league) has won another victory. As you do so, consider organized sports' new function. If Paul Tillich is right in defining religion as anything we take seriously, what might be our new national religion? Sports.

Sports fill the vacuum left by organized religion and sports fans (short for "fanatics") become pilgrims in our new crusades. Nationally televised sports play-offs, the World Series, the World Cup, the Super Bowl are generally better publicized and attended than most formal church events.

Instead of building new cathedrals, we build new stadiums and sports arenas, larger than ancient churches and tombs that once dominated the world. Athletes are not only among the highest-paid people in the world; they are also idols, held in godlike admiration, pursued by groupies.

Follow the money. Big sports = big business = big money. Get the new technology if you want to win—AstroTurf, springer pole-vaulting, in-helmet football radios, metal tennis rackets, computerized scouting—what's new this season? Get in the ball game!

The game itself is only the eye of the storm. The tidal waves are made by books, magazines, TV ads, films, endorsements. Only a limited number can sit in the stands—millions can watch on television, listen on radio, or read all about it in newspapers, magazines, and quickie books. Hundreds of hacks and journalists turn out thousands of words to snare you.

There are slick "specialty mags" such as *Sports Illustrated, Sporting News*, and scores of others centering on a single sport, team, player, or even a single game—hyped-up prose, pictures, graphs, tables, predictions. The leisure-time audience adds many millions to the go-get-'em fans in the stadium or arena. At home you can see the "magic moment" a second time—on instant replay. You can double your thrill. Instant replay suggests that time is recoverable, not lost in the flux of life. This in a sense makes the climactic moments immortal.

Sports thrills are much older than radio or television. Egypt and Greece had great sports fans. Bloody encounters of man versus the bull have been held for thousands of years in sacred places and still flourish. Such events were often rooted in religion; over the centuries those roots have remained. People in all epochs need myth, ceremony, and ritual to sustain them. This becomes their secular religion.

America has a genius for altering and inventing religions. Many abound today, especially in California. Others (including Mormonism) dominate large areas of America, and flourish abroad. But can any of these religions match the power and the glory of organized professional sports?

Once sports were mainly an amateur, gentlemen's pastime; but they are now a highly commercial competitive business for men and women of all ages. That story has often been told, and is well known.

Less attention has been paid to the people who migrated from the farms and country, or from the south to the urban north, to live in large impersonal cities. Sports offered a place where community and pride could develop, and ceremony flourish. They became the art of the people—the last heroic drama of a community. Popular culture wedded deep-felt needs and a new secular religion was born.

In the 1990s, in a world in which Capitalism finally brought down the Berlin Wall and Communism, and got the green light to push free enterprise and mass entertainment worldwide, American sports enjoyed unprecedented expansion and profit. Multimillion-dollar contracts for players who had only modest ability and multimillion-dollar facilities in cities that still had modest budgets for education, drug programs, and the ever-growing underclass dotted the map from coast to coast. Sports mania, especially where soccer is concerned, has infected the whole world. Whole cities are refurbished to accommodate the World Cup, held in Paris in 1998. In Latin America, soccer victories are a matter of life or death. The high priests and players of the sports world are both our saints and our sinners.

Sports has its Alphas and Omegas—beginnings and endings—winners and losers. Today winning isn't the main thing. It is (to quote football coach Red Sanders) the *only* thing. When the final whistle blows we know who has won, and who will be crowned with many crowns. We value this certainty. We don't find it often in our world—including our theology. This helps explain why some sociologists have said that sports have become the largest single waking activity of modern life.

* * *

We should consider just what sporting events mean to working-class people everywhere, as they go off to the stadium, spread out in a long line from pavement to pavement, eyes bright, yelling snatches of pop tunes. They are like children anticipating Christmas. When actuality lives up to imagination—the play, music, sport are as good as we hoped it would be, then we are fulfilled, satisfied.

What we describe in the twentieth century in connection with the world's most popular international sport (soccer) applies to other cultures stretching back centuries, even millennia. We know now that sports were popular in Egypt, and had strong religious overtones. The Olympic games of the Greeks (such as the Pythian, Isthmian, Nemean, and Athenaic) were sacred festivals, integral aspects of the religious life of the Hellenes. They were sacred games, staged in a sacred place and at a sacred festival—a religious act in honor of the deity. Those who took part, sports writers point out, did so to serve the gods. The Olympic games had their roots in religion. Over all the centuries, those roots have never been cut.

Look around you in contemporary America, if you want proof, as Super Sunday dawns. For centuries that happy morning was known as Easter, and the great hymn of Fortunatus (530-609 A.D.), "Welcome, Happy Morning," rang out over the land:

> "Welcome, happy morning!" age to age shall say.

The drama has moved from the empty tomb to the full stadium—not only in Christian countries, but everywhere. Games vary, but fans—quickly turned into fanatics—are a worldwide phenomenon as the twentieth century ends. What does this say about popular culture? And the nature of a possible global village?

Everyman (as John Bunyan called his Christian pilgrim) has become Everyfan. The metamorphosis is little studied or understood. Anyone can come up with qualitative judgments; little quantitative information is available. Even a stopwatch, in the hand of Professor Michael Real, can provide new insights. He found there was surprisingly little action in TV's "action-filled" bowl games. Pre- and postgame activities consume 20 percent of the TV time, advertising another 15 percent, comments, between-play, and half time another 40 percent. Much of the remaining 25 percent "scoreboard clock time" goes for huddles, decisions, and penalties. In fact, during the four-hour telecast, there is *less than ten minutes* when the football is actually in motion.

What Professor Real does is analyze the mythic functions of media sports (using the Super Bowl as a model), showing how events of no real consequence are raised to powerful, quasi-sacred myth—larger-than-life drama. He finds that media myths function in a manner quite similar to traditional mythic attitudes, providing heroic archetypes, personal fame, and spatial and temporal frames of reference. So, like major sporting events in all ages and cultures, the Super Bowl is a communal holiday and display of cultural values.

This was especially true in ancient Rome, where the spoils of imperialism were lavishly celebrated. We still use the Roman word coliseum (from the Latin *colosseum,* colossal, huge) for our largest sports stadiums. The Super Bowl is a mythic prototype of our ideas, celebrated collectively. So, on a global scale, is the World Cup.

Super sports events impose on the players complex and coordinated demands that exist in the world outside the stadium—in assembly lines, tract homes, superhighways, and mastery of computer technology.

But these things, one might point out, are not religious but secular, even commercial. In today's world, who can separate them, and how? Do we live, not in the City of God, but in the Secular City?

* * *

Consider these religion and sports similarities. Both have sacred songs, places of gathering, favored animals—eagles, doves, lions, bulls. Both use incantations, and depend on chants, cheers, and songs. Instead of knights seeking the Holy Grail, we have players seeking the Super Bowl gold ring, the Olympic gold medal, or the World Cup. They all create a special atmosphere, which results in excitement, even frenzy.

If the church people have prayers, the sports people have yells. Countless imitations of chants with tripartite patterns flourish. Here is an early one that D. A. Rollins introduced to Dartmouth in 1878:

> Wah-hoo-WAH! Wah-hoo-WAH!
> Da-Da-Dartmouth, Wah-hoo-WAH!

Yale fans got their "Brekekekex-koas, koax" from Aristophanes' "Frog Chorus," and have been croaking away since 1884. Fans have their saints (or stars) such as Jackie Robinson, Arthur Ashe, Babe Ruth, Ted Williams, Tiger Woods, and Michael Jordan.

Professional athletes do not play for fun; they play to win. The more they win, the more they make. They are not engaged in friendly games, but minutely observed and monitored battles in which the winner takes all. Isn't this the way it is in the corporate world? Are sports not a mirror image of Life Out There?

Perhaps also of Life Up There? Some coaches insist on prayers in the locker room. Picture prayers by professional football teams. Giant men, bruised and sweating, kneeling to beseech the Almighty, the Divine Coach, to let them score big. Muscles, guts, and good luck make all the difference. The struggle is primordial. In big-time football, the most popular sport, you can often see real blood—not makeup, as in the movies or on television. Real blood. Real pain.

The world seems to need super sports events, in which sweat and blood mix. Every fan knows and expects this.

Super Sunday had been super—and after several dull years, was super again in 1999. To quote headlines in the February 1, 1999 *Washington Post:* the "Denver Broncos Ran Wild," to "Corral the Falcons, 34-19." Capping a sixteen-year career, John Elway "Added Another Chapter to Legend."

Thus did Super Bowl XXXIII hold its 140,000,000 viewers, and justify a host of advertisers who paid $1.2 million for a thirty-second ad shown during breaks. Fans could go home tired but satisfied. Technicians could rewind miles of cable and wire. Editors could scan more miles of films, made from scores of cameras, and isolate "the right shots." Instant replay!

The hype had worked, and it will work again. The nation will keep alive the memory, the liturgy, the "spirit." Personal trials come and go, but the best moments inside church, arena, and stadium live on and on. "You remember when late in the fourth quarter . . . " O come, O come, Immanuel . . .

* * *

ENDNOTE

Popular books, magazines, and films about both religion and sports come out in an endless stream. The Further Reading section in my *Great Awakenings: Popular Religion and Popular Culture* (Binghamton, NY, The Haworth Press, 1995) should be helpful. The two standard scholarly volumes are Sydney E. Ahlstrom, *A Religious History of the American People*, 2 vols. (New York, Doubleday, 1972); and Martin Marty, *Pilgrims in Their Own Land: 500 Years of American Religion* (New York, Little, Brown, 1984).

The classic books about sports genesis are Jean Le Plochmoan, *La Genese des Sports* (Paris, Payot, 1962) and Fred Inglis, *The Name of the Game: Sports and Society* (London, Heinemann, 1977). For the union of sports and popular culture, see Michael R. Real, *Mass-Mediated Culture* (Englewood Cliffs, NJ, Prentice Hall, 1977) and John M. Carroll, *Sports in Modern America* (St. Louis, River City Publishers, 1981).

These two recent titles point in new directions:

1. Genevieve Rail, ed., *Sport and Postmodern Times* (Albany, NY, State University of New York Press, 1998).
2. Phyllis J. Perry, *Exploring the World of Sports: Linking Fiction to Nonfiction.* (Englewood, CO, Libraries Unlimited, 1998).

Ray and Ronald

Bill Clinton was reelected president of the United States in 1996, promising to "build a bridge to the future." One visible and viable American bridge had already been built, reaching every corner of the globe. Its symbol was not the American flag but the Golden Arches. That bridge was McDonald's; today the term "McDonaldization" is used to characterize similar changes and innovations in many other enterprises.

McDonald's is, to the casual observer, a chain of highly successful restaurants specializing in hamburgers (or in India, lamburgers) and french fries. In a broader sense, it is a macrocosm of America and its all-pervasive popular culture. Perhaps, as anthropologist Conrad Phillip Kottak maintains, it is even a secular religion, its restaurants offering a kind of spirituality without theological doctrine.

As our century ends and a new millennium begins, the impact of fast foods and highly rationalized operations has had a major impact not only on our stomachs but on our psyches and lifestyles. Events worldwide bear this out. Once, according to an old platitude, all roads led to Rome. Now some would say they lead to McDonald's. Holy Hamburger!

Who made this possible? A remarkable salesman entrepreneur named Ray Kroc, and the clown (Ronald McDonald) created as his spokesman. Ronald loves hamburgers and children, and works magic. Part of that magic has been to make a large number of people millionaires. O brave new world—The world of Ronald McDonald.[1]

The McDonald's saga has become public property.[2] Ray Kroc himself got into the act, by writing a brief autobiography.[3] The spectacular success of McDonald's in the global village of the 1990s (especially in Russia and China) occurred at a time when the world was full of spectacular failures. In 1826, Jean Brillat-Savarin observed that "The destiny of nations depends on the manner in which they nourish themselves." More and more we nourish ourselves on fast food. How have Ray and Ronald done it?

One can best answer by reviewing the early days of McDonald's, and the triumphs on the home front—for the formula has been little changed in other countries. The huge emphasis on localizing the architecture and hiring native managers and employees tends to conceal this fact.

Our story begins in 1930, when Richard and Maurice (widely known as Dick and Mac) McDonald left their native New Hampshire in 1930 and moved to California to make their fortune. After several false starts, they decided in 1937 to open a drive-in restaurant just east of Pasadena. Dick and Mac cooked the hot dogs (not hamburgers), mixed the shakes, and waited on customers seated on a dozen stools. Three carhops took care of the parking lot.

By 1940, the enterprising brothers had made enough money to open a bigger drive-in fifty miles east of Los Angeles, in San Bernardino. By 1945 it was the town's chief teenage hangout, employing twenty carhops. But they saw they could go no further without a new format. In 1948 they closed, fired the twenty carhops, redesigned their restaurant, and put new emphasis on speed and volume production. The menu was slashed from twenty-five items to nine, the price of a hamburger from 30 cents to 15 cents. This was the restaurant that Ray Kroc entered when he went west to sell milkshake mixers to the McDonalds. Instead, he ended up buying their idea, their franchise in 1955, and eventually their name for $2.7 million in 1962.[4]

In 1955, Kroc opened up a flashy structure, in Des Plaines, Illinois, featuring golden arches designed earlier for the McDonald brothers by Stanley Clark Meston, a pragmatic problem solver rather than a modern architect. He wanted something that could be seen. There was for him no symbolic or historical association. That same label—pragmatic problem solver—has remained the key to McDonald's growth.

Meston's logical solution to a functional requirement—catching the attention of speeding motorists—was two twenty-five-foot yellow sheet-metal arches rising through a flaring wedge-shaped stucco roof. The arches were signs, self-supporting and not part of the building's structure. They have become one of the most recognized signs in the world, embodying changes in American lifestyles and, later on, throughout the world.[5]

The yellow arches atop the red and white tiled restaurant were prominent fixtures on the American roadside landscape from 1955 onward, finding a permanent spot in the American psyche. Over 1,000 facsimiles of this prototype helped to establish the fast-food industry before television took over as the primary medium for mass promotion. In 1993 at the prestigious International Fair of Technology in Paris a replica of that first McDonald's, resplendent with its golden arches, was on view for millions of admirers. The world had been McDonaldized. That same year more than half of McDonald's profits came from abroad.

The 1955 prototype restaurant was modified to accommodate different climates and expanded service. Basements were added for furnaces and storage; the service area was enclosed in glass walls. The sheet-metal arches were replaced with back-lighted plastic arches. During the 1960s the arches, originally designed as parabolas, were rounded off. The newly designed restaurants featured a low-profile mansard roof with brick and shingle texture. But the golden arches—popular commercial vernacular style—had taken on a symbolic life of their own.[6]

So had Ronald McDonald. Clowns are attention getters, the source of fun, laughter, and trickery. The official spokesman for the McDonald's Corporation, Ronald is the world's best-publicized clown. He dances, prances, and becomes one with children. He works magic, loves hamburgers, and defies the laws of gravity. Ronald McDonald Houses host families whose children are hospitalized. He is the very essence of goodwill.

The first Ronald made his local debut in Washington, and his national debut a few months later in Macy's Thanksgiving Day Parade—New York, November 25, 1963. He went on television in 1966, and was dubbed "official spokesman" in 1967. Over the years, by constant exposure, he has become part of our media landscape: the genial host of McDonaldland.

What does Ronald do? His "official biography," sponsored by the corporation, tells us: He spends all his time going from one McDonald's restaurant to another to see his friends, the children. Suffer the little children to come unto me.

And what does Ray do? Figure out how to get more and more customers to come unto him. Always a nontraditionalist, he gathered together a group of executives, many of whom had no college education. Kroc was not interested in theories, but in sales. His people should have what he called "real-world degrees."

It was a bit ironic that they decided to open a university—Hamburger University—to train franchisers, staff, and managers. It turned out its first graduating class of fifteen in February 1961. *Life* magazine and the *CBS Evening News* took note. So did academics, appalled at the idea of a degree in Hamburgerology. But Kroc's university flourished, moving into a $40 million facility and offering thirty-six hours of courses accredited by the American Council of Education.

Success breeds success—keep growing, Ray insisted. For architects, designing fast-food restaurants is a challenge. "It's like positioning a dining room in front of a factory," architect David Norris complained. The goal was not a machine for living (recall Le Corbusier's famous definition of a house) but a machine for eating. How to balance graphics and garbage disposal? How to blend art and commercialism? Aesthetics and profit? How to keep the home town atmosphere and friendliness? Kroc appointed a top executive, James S. Kuhn, whose job was "to make the company feel small." He was Vice President for Individuality.

Ray quickly expanded his empire. He bought an upscale Chicago restaurant and a chain of pie shops and came close to buying Taco Bell and Baskin Robbins. Then his thoughts went international. To introduce the new Quarter Pounder, the company held its First International Convention in Hawaii in 1972. Three thousand came to that island paradise to savor the new concoction—managers, operators, top-level executives. Ray Kroc flew in on his private jet. There were Broadway-style songs and dances and myriads of Big Macs.

If he were to succeed in new countries, there must be major cultural experiments. Fast food and drive-ins were unique to the United States. For a while going elsewhere was blind man's bluff. Stores in the Caribbean and Canada ran into unexpected problems. In Europe the first operation, in the Netherlands, was a joint venture with the country's leading supermarket chain, Albert Heijn. There were miscalculations. The restaurants were built in the suburbs, instead of the thriving cities. To please the Dutch, applesauce and deep-fried chicken croquettes were added to the menu. Perennial losses followed.

The big breakthrough came in Japan, where that country's Ray Kroc—Den Fujita—got into the act with a joint venture. He refused to alter the menu—he would stand or fall with the American hamburger. Always a master of press coverage and controversy, Fujita had this to say:

> The reason Japanese people are so short and have yellow skins is because they have eaten nothing but fish and rice for 2,000 years. If we eat McDonald's hamburgers and potatoes for a thousand years, we will become taller, our skin will become white, and our hair blonde.[7]

Not only did Fujita succeed—there are now over a thousand McDonald's in Japan—but he set the basic philosophy of all international ventures that followed:

- When entering a country choose the most visible and commercial downtown spots, no matter what the cost.
- Use local real estate agents, architects, and contractors for the building.
- Stick to the basic three on the menu—hamburgers, fries, and shakes; insist on efficiency, cleanliness, and friendliness.
- Stay in for the long run—never give up. Remember the sign on the wall of Hamburger University: PERSISTENCE AND DETERMINATION ALONE ARE OMNIPOTENT.

Country after country came eagerly and profitably under the golden arches—Western Europe, Eastern Europe, Australia, and Asia. The image and the icon were being transferred with the business; there were social, cultural, and economic ramifications. By 1993 the international restaurants were more profitable than those at home. The most successful countries have been Japan, England, Canada, Australia, Germany, and Austria. These six now account for 80 percent of the overseas revenue. Great use is made of advertising, especially on television and in public places. The last time I was in London in the Underground stations there were large posters of the city's most famous sights—all giving the distance of each from the nearest McDonald's.

Kroc hasn't hesitated to go into third world countries where the average annual income is only a few hundred dollars, and a Big Mac is a luxury. Especially promising in recent years are Brazil, Korea, and Indonesia.

One of the major reasons for success has been architecture. Brad Trask, who deals with international matters for McDonald's, explains:

> We take great care in the design of our restaurants to blend into the surrounding environment. Our locations at the Spanish Steps in Rome or on the High Street in Hampstead, United Kingdom are virtually indistinguishable from surrounding buildings. Key to our ability to achieve this is the use of local architectural firms, designers, and contractors.[8]

Flooring, seating, and decorating schemes often reflect local tastes and cultures. The McDonald's on Syntagma Square in Athens, for example, uses two-story marble columns, courtyard fountains, and wall murals by local artists. In Beijing modern styles are combined with traditional touches, such as the round, red, Chinese-style windows.

Sometimes old buildings are renovated—as in São Paulo, Brazil, or Bucharest, Romania, where the old railroad station became a new restaurant. In other countries, as our illustrations show, both the architecture and the spirit of the country are evoked. In all cases, local people are highly visible. The effort is to make every McDonald's look local. Only a few dozen American officials have been posted abroad, and every effort is made to favor the indigenous.

But make no mistake about it—there is nothing indigenous behind the counter. Brad Trask explains:

> Our kitchen designs and equipment are standard and finely honed with years of experience and attention. As a food business, the kitchen is the heart of our business and our formula varies little. This allows the McDonald's experience in Paris, Texas to match that of Paris, France.[9]

This seems to be the formula: the exterior belongs to them; the interior design is part theirs and part ours; but the work area and kitchen are ours.

The two greatest triumphs abroad have come in Russia and China. In Moscow's Pushkin Square, amidst the drab socialist architecture and dingy gray buildings sits the new McDonald's—bright lights, glass walls, gleaming plastic—a gingerbread house in an army camp. If *glasnost* has a monument, it's this restaurant. To handle the opening Bolshoi Mak attack in February 1990, which had required fourteen years of agonizing negotiation, nothing was spared. The restaurant had a seating capacity of 700 and could serve up to 15,000 customers a day. On opening day thousands of Muscovites poured in to sip "milk cocktails" and taste their first Bolshoi Mak, picking them apart to marvel at the "fixin's." A burger, fries, and cola cost half a day's wages for most customers; but they are willing to stand in line for hours to buy them.

Two years later, the world's largest McDonald's opened on the busiest intersection in the capital of the world's most populous country. Not even the Great Wall of China could keep Ray and Ronald out of Beijing.

When the doors opened at 8:35 a.m. on April 23, 1992, the 700 seats were instantly filled as people climbed over metal barriers, knocking into people standing in line. Maidanglao, as McDonald's is called in Chinese, was a sensation. Big Mac became Juwuba—"big with no equal." Located not far from Tiananmen Square and the tomb of Mao Tse-tung, how could it be otherwise?

"Our research shows that 800,000 people pass by our corner every day," said James R. Cantalupo, president of McDonald's International. He also pointed out that Beijing has a population of 10.3 million, and thousands of other Chinese pour into the city each day on business or as tourists. The company hopes to open twenty more restaurants in China, investing $50 million.

Now McDonald's is eyeing what might be its most difficult conquest—India, where four out of five people are Hindus who cannot eat beef. If not a hamburger, how about a lamburger? Or a new menu that appeals to Hindus? India, which in 1977 ejected many foreign companies as imperialists, has changed its stance. Coca-Cola and PepsiCo are everywhere, and in 1993 McDonald's was granted permission to open twenty stores. Deepak Nirula, president of India's largest fast-food chain, is pleased.

How can McDonald's pass by a country with 936 million people—one out of every six people in the world—whose middle class is larger than the population of the United States? Big Mac, launch your attack! But beware of attacks on the home front.

As Ray and Ronald's fame and fortune grew, so did the arguments of their detractors. Sociologist Amitai Etzioni held that the restaurants were unfit for young people: These are breeding grounds for robots working for yesterday's assembly lines, not tomorrow's high-tech posts. Are we creating a Brave New World under the golden arches, and exporting it from Berlin to Bangkok, Texas to Tokyo? Observe kimono-clad Japanese women leaving Shinto shrines going to "Makudonarudo" to enjoy a Biggu Makku. There, and throughout the world, burgers sizzle, shakes shake, Big Macs slide down stainless steel ramps, and electronic cash registers beep. Is the whole world being McDonaldized?

That was the contention of George Ritzer, a professor of sociology at the University of Maryland, whose book titled *The McDonaldization of Society* appeared in 1995. Ritzer defines McDonaldization as the process by which the principles of the fast-food restaurant are coming to dominate more and more sectors of society. This is a control revolution. Control is ageless, but what he calls "the revolution" is comparatively recent: photography and telegraphy (1830s), power printing (1840s), typewriter (1860s), telephone (1876), moving pictures (1894), radio (1895), assembly-line production (1903), television (1923), and the computer (1944). All this gave rise to what Max Weber called the ultimate control—bureaucracy. And the ultimate benefactor is McDonald's.

Behind Ray Kroc stands Henry Ford and Fordism—the mass production of multiples of identical products from readily available materials. This increases the quantity and decreases the quality. But there are advantages to the fast-food restaurant, as well as other elements of a McDonaldized society. People now have easier access to other foods—Cajun, Chinese, Italian, and Vietnamese, for example. Salad bars provide healthier food, made exactly the way the customer wants. Malls are comfortable and efficient, and home shopping networks allow those who wish it to shop without leaving home. Automated bank tellers provide money any time of the day or night. Considering these and other matters, George Ritzer asks some profound questions: Is McDonaldization the equivalent of modernity? What *isn't* McDonaldized?[10]

The basic premise is that McDonaldization is much more than the story of how to produce hamburgers and french fries. It's a way of looking at the world with its new types of consumption—supermalls, superstores, TV shopping networks, cybermalls, and infomercials.

McDonaldization has taken on significance that earlier inventors and theorists could not have even imagined. The land of the free is the land of the "Mc." We have McDentists, McDoctors, McChild Centers. *USA Today* has become our McPaper, serving us News McNuggets. The McModel includes Jiffy-Lube, H&R Block, Pearle Vision Centers, AAMCO, AT&T, and Kinder Care (also known as "Kentucky Fried Children"). And isn't that model working for many of the purveyors of pop McReligion?

There are striking parallels between the rise of Big (bionic) Mac and Big (electronic) Jesus. The first McDonald's opened in 1955, when Billy Graham's ministry was making national news, and television was becoming evangelism's medium of choice. The goal was products—customers for hamburgers, converts for evangelists. The formula had been spelled out in the 1920s by the social theorist Max Weber: standardize, mechanize, simplify. This would increase efficiency and productivity but, Weber warned, it would diminish the individual. "The system" and the bureaucrats would take over. In some places they already have.

The key word is *fast:* not only fast food but fast computers, conversion, music, information, disinformation. Marshall McLuhan seems to have been right: if it works, it's obsolete.

Certainly it has worked for McDonald's. The corporation now has 14,000 restaurants, and plans to open 1,000 new ones a year. No other system in the world can offer such efficiency, economy, predictability, productivity. And "They do it all for you." This might be translated, in the world of popular religion, as Amazing Grace.

Big Mac and Big Jesus have much in common. Both offer a product, disguised as entertainment. Fast-food restaurants are an amusement park for food; televangelists' broadcasts are an amusement park for religion. Both have color, phony fellowship, and cliché blessings. ("Have a nice day." "The Lord bless you.") Young people, with sparkling teeth, help set the tone. They are part of public theater rather than private dining or worship. They set the stage for a superbly orchestrated sales campaign.

The New Hucksters sell hamburgers or salvation with a program based on the three Ps: promotion, packaging, and profit. Clearly they have sold billions of hamburgers; only God knows how many souls have been saved. The earthly claims run into the millions.

Though the radio ministry seems to be diminishing, it is still a potent force—and has been since the 1930s when Father Charles E. Coughlin, "The Radio Priest," employed 106 clerks and four secretaries to handle his incoming mail. His strategy was not to inform but to inflame. He denounced Jews and Franklin D. Roosevelt. His Sunday Talks became a weekly ritual for millions and his church in Royal Oak, Michigan became a shrine. That was B.C.—before computers. Now computer chip persuasion has become a new art form.

All sorts of people have joined in for what some critics have called "The rape of the vulnerable." Not only the televangelists but also the Moonies, Jesus freaks, body mystics, gurus, Branch Davidians, hands-on healers, out-of-body tourists, speakers-in-tongues, snake handlers, willful witches, and trembling exorcists have gotten into the act. What three-ring circus could match it?

When religion becomes a multimedia sales campaign, conversion may become a pseudoevent.

As the century ends, McDonald's itself faces new problems. So fierce is the competition that McDonald's has been losing part of its market share. Introducing a new line of "deluxe" foods in 1996 didn't seem to help. There is also the problem of controlling and administering so vast and expanding an empire. Ray Kroc, like Sam Walton, who founded Wal-Mart, had the personal touch and charisma that seemed to disappear when he died. Did he carry the key to his success to the grave? Are we backing away from a completely rationalized society?

The people who run McDonald's now don't think so. Ray Kroc, they say, is in the grave—but his soul goes marching on.

PART V: AFTERTHOUGHTS

The Iceman Cometh—Again

Alpha and Omega—the beginning and the end of our journey takes us back thousands of years to the dim primordial past.

We first encountered popular culture by peering into the dark museum home of Oetzi the Iceman, who died some 5,300 years ago in the Alpine glaciers; then into ancient caves, where pictures of giant animals skillfully depicted the struggle for survival. We wondered what sort of people our primitive ancestors were, as they moved from caves to tombs to forums to castles to cities. Oetzi, the best-preserved Bronze Age man ever found, was discovered in the 1990s in the high Dolomite Mountains. He brought the past into the present. Fully clothed and armed, he was ready for the hunt.

Like this remarkable mummy, we have been on our own hunt, going through periods of time and major themes—"the people," style, symbols, lores, fads, falling stars, final thoughts.

We have attempted to explore links that illuminate our journey, such as myths and dreams, East and West. Everyman and Everyfan. Various aspects of cyberspace have been examined and challenged.

What are the parameters of popular culture, and how are they changing in the Information Age? Are we moving forward to a new globalism or backward to the old tribalism? What will life be like in the newly formed cyburbias? Are they yet another utopian dream?

Employing new techniques, theories, and metaphors, can we begin to make an aerial mosaic that gives an overall picture of the remarkable human saga? We hope to have made that beginning. Now let others go on to lead to new understandings. The road ahead lies open.

We Are One—and Yet . . .

Despite our constant conflicts, wars, and hatred, we are one—we are all members of a single species, Homo sapiens. Multiculturalism notwithstanding, we are much more alike than different. Nine-tenths of us as adults measure from four feet ten inches to six feet and weigh from 90 to 220 pounds. We require 2,500 to 4,000 calories per day to be strong and healthy; we can manipulate our thumbs, conceptualize thoughts, and speak as no other animals can. We can store knowledge and make myths. We can modify our instincts, acquire neuroses, and mourn our dead.

Just how we acquired all these skills, and how we have evolved over millions of years, remains something of a mystery. How popular culture develops, functions, and changes is a part of that mystery.

While we are one—*unum*—we are also many—*pluribus. E pluribus unum*—out of many, one, is our American motto, our noble hope, but we all know it is far from a daily reality.

Our special circumstance, as the twentieth century ends, is that we are all so interconnected and intertwined that things happening far away can alter our lives and our future. Many examples come quickly to mind. Crises in the Middle East affect oil prices everywhere. New York gets the Hong Kong flu. The Russians build a series of dams in Egypt, the Japanese a hydroelectric system in Sri Lanka. The Italians build Fiats for the Soviet Union, the Americans Fords for Brazil. American troops wage war on the rim of Asia, Finns guard the border between Israel and Egypt, United Nations troops patrol Bosnia. The sudden collapse of Asian economies sends shock waves throughout the world's economy. Wall Street sneezes and Europe catches a cold.

On this level, globalism has become a fact of life. But times change. The Russians have left Egypt, the Americans Vietnam, and everyone wants to leave Bosnia. All the old tensions remain, and all the cultural traits and preferences remain in place. Some of them even become more virulent and popular. The tendency to defend one's local turf is a real universal principle at a time when political credos appear, flourish, then fall into oblivion.

Monoculture

The word "globalism" has a peppy progressive sound. Calling it monoculture casts it in a different light. Now it sounds a lot like monopoly, monochrome, and monotonous.

Want it or not, monoculture is coming, some say; and we can do nothing to avoid it. We have electrified and massified not only our technology but our social institutions and private psyches. People are the medium, money is the message. The new monoculture has become a monoindustry; advancing under the banner of consumer capitalism, on-line onslaughts, and deregulated grabbing.

Overwhelmed by American technology, military power, and mass entertainment, some fear the world must play the game by American rules—umpired exclusively by Americans.

No nation resists this more than the French, who have been a big loser in the twentieth-century power transformation. They are not like us—and they don't want to be. Their intellectuals see mass culture, and not religion, as the opiate of the people. They are overwhelmed and undermined by our movies, television programs, advertising jingles, sound bites, yellow journalism, and incomprehensible rock and rap song lyrics—say their intellectuals. Ordinary people, on the other hand, find all these things fascinating and irresistible.

How best to catch the "flavor" or "tone" of America at the century's end? There must be a hundred ways, and none would really do the job. In today's business-is-king culture, during the greatest boom of the century, why not check the best-sellers in the business area? Why are people spending millions of dollars on popular books to fulfill their dreams and line their pockets? These ten top sellers say a great deal about the flavor of the 1990s:

1. *The Seven Habits of Highly Effective People* (Habitually popular. Number one on the list for many weeks. Think success!)

2. *Success Is a Choice* (Coach says harder, harder!)
3. *Only the Paranoid Survive* (How to turn, overcome crises)
4. *Built to Last* (How Wal-Mart and others became standouts)
5. *Stock Market Miracles* (Infomercials for the author)
6. *Selling the Invisible* (A collection of bite-size lessons)
7. *Wall Street Money Machine* (A cab driver gets rich)
8. *Get a Financial Life* (Hey, Gen-Xers—money means fun!)
9. *Investment Guide* (Recipe for 23 percent returns. Big bucks!)
10. *Against the Gods* (Try risk management—the big secret)

Many European intellectuals and academics regard culture as a public service and trust—not a chance to make money and entrance the young. In America, they say, culture is just another product, perpetually for sale; no different from detergents and deodorants.

Hence the lines between high, low, folk, and pop culture are continually blurred, even eliminated in America. Europeans often agree with American critics such as Dwight Macdonald, who detested American vulgarity and kitsch, which, he said, diluted and threatened to eliminate regional cultures, crafts, and class traditions.

The streamlined slick glamour of American culture didn't stay at home. It crossed oceans and borders, bent on creating a monoculture. Barbarism was at the world's gates, and the gatekeepers were being overrun. Big Mac, the new Genghis Khan, would be victorious.

Today's take-all technocrats hold that the traditions, pleasures, and lifestyles of the past must be sacrificed for a more productive technological future. This argument, Jacques Ellul contends (in *The Technological Society,* 1967), creates the "modern myths" of work, progress, and happiness; and what Neil Postman calls (in *Technopoly,* 1992), "the deification of technology."

As the great Information Revolution moves forward ever faster, the Luddites are voices in the wilderness. May I make one prediction about the unpredictable postmodern world? Luddites will grow ever stronger, and like other crusades that seemed "misguided" not many years ago (against nuclear power, tobacco, unregulated ocean fishing, patriarchal domination, etc.) theirs will assume new power.

This will be no easy task. When myths are filling the stomachs and pocketbooks of millions—without asking about other aspects of life—the faithless doubters have an uphill struggle. Big money, big power, big propaganda are in the hands of the mighty. They try to persuade us that their techno-toys give access to Divine Omniscience, stemming from a mystical Celestial Library. Bill Gates and his Merry Pranksters hold the key to the kingdom.

There is a flaw in their heaven. In its sacral quest toward a better life, it casts thousands into the pit of unemployment, and transforms others into disposable objects. Silicon Valley may be seen in a different light—as the newest version of the sweatshop. A whole series of movies depicting disenchanted and largely abandoned children—*Reality Bites, Kid, Mallrats, The Doom Generation, Pump Up the Volume*—show that all is not well on the home front. The myth of progress is being replaced by malaise. The same message emerges in the popular comic strip *Dilbert*, where futile work consumes people who see life mainly through pictures on the World Wide Web. We must offer a fresh alternative to all this.

Variety and diversity are so deeply rooted in human life that monoculture, despite temporary triumphs, will always be rejected. The people will right the ship of state, as they always have.

Fling Wide the Gates!

FLING WIDE THE GATES, LET THE MILLENNIUM IN! READY OR NOT, IT IS ALMOST HERE!

No everyday event this—it occurs once every thousand years. No one living has experienced one, and no one alive now will live to see another. Reactions are mixed: hope versus fear, utopia versus dystopia, heaven versus hell. All this is reflected in our popular culture—the mirror held up to the culture.

Trauma, a buzzword of the 1990s, has been buzzing overtime. The word, derived from the Greek *traumatos*, means a wound; the Greeks made much of trauma in their drama, philosophy, and thinking. So do we.

Trauma . . . an injury or wound violently produced—the condition or neurosis resulting from this. Trauma was central to the thinking of Sigmund Freud, who is central to any history of the twentieth century. He discovered and stressed that strong emotional experiences or shocks have a lasting psychic impact. Look at what has happened in our times: no wonder we are traumatized.

In addition to all the usual problems, there are new ones: the uncertain effects of the new technology, the hazards involved in the new globalization, the political disarray after the first presidential impeachment in over a century. Three words echo through all aspects of a fast-changing society: race, gender, and ethnicity. We must deal with them all, and many other matters that will emerge.

As the century ends, and much of the world struggles with war, disease, and economic depression, the United States appears to be the only candidate for world power—perhaps the only power in history to have the means for empire without the overweening desire for it. Instead of conquering with bullets and bombs, we may do so with our popular culture—films, videos, tapes, books, and advertisements.

The idea of this kind of empire appeals to many thinkers and historians around the world. One of these, Hugh Thomas, confirms this on the final page of his *World History: The Story of Mankind from Prehistory to the Present:*

> A world dominated by a single, restrained yet capable great power would be a far safer place than existed between 1945 and 1989. It would probably be a better world than any other which can reasonably be imagined.[1]

If there is a chance to create and sustain such a world, then let us work hard to fling wide the gates.

The Pop Heard Round the World

Here once the embattled farmers stood
And fired the shot heard round the world.

Ralph Waldo Emerson

The embattled farmer who launched the American Revolution is long dead, turned into a bronze statue. His "shot heard round the world" came from a primitive rifle. In his place is the triumphant media mogul. He uses not a rifle but the newest high-tech devices with deadly accuracy. His ammunition is American popular culture. His is the new pop heard round the world.

Our biggest export comes not from our fields or factories, but from our soft drinks, fast-food outlets, movies, TV programs, music, and computer software. While intellectuals debate the virtues and devices of our popular culture, the world yearns for and embraces them. Travel the seven seas, and what awaits you? A Coke bottle, a Big Mac, an episode from *The Young and the Restless,* or a pirated showing of *Titanic.*

What are the young people doing on the street corner? Following the advice of the Marlboro Man, and lighting up another cigarette. Stand alongside and watch the Fords go by.

What does all of this mean? A giant power shift has occurred in our century. Ancient Persia and Sparta were military states, Israel and Egypt religious states, modern England and Japan industrial states. Now the United States has become the world's first electronic media state, conquering with images, entertainment, and virtual reality.

Rome controlled her empire by force, the papacy and Catholicism by prayer and devotion, the British with their navy and diplomacy, the Marxist-Leninists by ideology. Now it is our turn. The media, which we dominate worldwide, is the message, the massage, and the money. The world is on-line: wired for sight and sound. What sort of world awaits us?[1]

International sales of our software, films, books, and entertainment products totaled $60.2 billion in 1996—more than any other U.S. product or industry. Since 1991, when the Soviet Union collapsed and new world free markets sprang up, total exports of intellectual property from the United States have risen 94 percent. No one knows how many additional billions of revenue dollars were lost to pirating and illegal copying.

A new international style—powered by atomic energy, monitored by satellite, grounded in popular culture—has transformed the world. Replacing earlier styles (classic, gothic, baroque, modern), it has leaped over old political, social, linguistic, gender, and racial barriers. It links New York with New Delhi, Paris with Peoria, Roanoke with Russia, Bali with Beijing. We have altered the product of many centuries in the twinkling of an eye.

Behind figures is a pattern, perhaps a cultural revolution. Pop goes the world.

The Triumph of Pop—but just what *is* it? The dictionary lists over twenty definitions of the verb "pop"—to say nothing of the noun, adjective, abbreviations, and link words (including popcorn, pop-eyed, popgun, etc.). Usually "pop" means one of three things today: (1) new, faddish, "in"; (2) vernacular, folksy, earthy; or (3) universal, electronic, instant. All point to an unflinching look at today's baffling buffeted world—mechanized, trivialized, advertised—overflowing with newness and nowness, hype and hope, abundance and anxiety.

The nineteenth century took its toll on workers' bodies; the twentieth, on their nerves and minds. We fight time, and often lose the battle. "Instant" is the key—instant coffee, burgers, soup, sex. Who has time to waste on preliminaries these days? Besides, you can always switch instantly to another brand. Perhaps Carl Sandburg was right: the past is a bucket of ashes. "I'll never forget old what's-his-name."

But Americans must never forget that many ancient well-rooted cultures (such as those of China, Iran, Israel, France) don't want to forget the past, and will never call it a "bucket of ashes." They inherit old mythologies, and are faced with hostile new ones championed by our films, magazines, television, and computers. How they will mix with the older, more comprehensive myths no one can say.

Will the "American intrusion" increase or decrease hostilities? Will our unprecedented prosperity spread to other lands—or will they fall into deep depression (as is already the case in some places)? How much will our popular culture become part of their own? We may have the technology, but others have their own ideologies, philosophies, and social theories. For them, things might *not* go better with Coke. They might not want their daughters to grow up to imitate our Madonna.

These are exciting but perilous times. The role international popular culture will play is far from certain. We have wired the world for sight and sound—but how does it sound and look to all the people who are wired? Who will monitor and regulate the new technologies—not only on earth, but orbiting the earth and cluttering up the heavens?

The old nationalism may not be replaced by globalism but tribalism, with vicious and bloody outbreaks which plague our world today. Might our technology and popular culture, combined with the English language, point to a new imperialism? As we get ever richer, will much of the world become poorer, not only economically but spiritually and culturally as well?

These and many other vexing questions will demand all the skill, knowledge, patience, and vision we can muster. A new millennium, and a new world, bring new challenges.

Some Final Thoughts

In 1508 the great Renaissance artist Michelangelo was called to Rome by the pope to paint the ceiling of the Sistine Chapel. It was a staggering task for a man who was primarily a sculptor, and took years to complete. The result: one of the most famous art shrines in the Western world, and images that seem to grow even more powerful and relevant as the centuries pass.

Michelangelo set out, as part of his task, to retell the story of the Book of Genesis in the Bible: the creation of man and woman, their fall and expulsion from the Garden of Eden, and their agony as they faced the wilderness of the world with its continuation of sin.

God reaches out, in this famous depiction, to touch man, whom he wants to "create in his own image." Both the hands of God and man reach out—but do not touch. There is a gap between their two forefingers. This vivid depiction has been used innumerable times to express such things as heroic struggles, false expectations, and tragic failure to be what God had first intended us to be. Time and again, in our own day, we see the image of God reaching out to touch (and sell) books, computers, pencils, television sets, furniture—not only to endorse them but somehow to grant them divine favor. The fact that this might well seem to earlier ages a form of blasphemy doesn't bother today's admakers and hucksters. After all, it works.

Sometimes we are shown two people—say, politicians, athletes, or candidates for election, "reaching out"—again, seeking in some bizarre way God's endorsement. One of the strangest ads, which my students always puzzle over, shows a young woman standing on a toilet, painting Michelangelo's "Creation of Man" on her bathroom wall. Who sponsors the ad? Drum chewing tobacco. What is the connection?

The Man-God gap catches our imagination when applied to any number of things—gender gap, racial gap, income gap, language gap, age gap. At the moment, it highlights the tragic splits in Congress and our national agenda that have caused great anger and bitterness. President Lyndon Johnson urged us to "Come sit and reason together." We seem instead to sit and squabble.

We talk peace—but wars rage. Over a dozen nasty struggles continue year after year, in Europe, Asia, and Africa. We condemn and threaten—but do little or nothing to close the gap. Genocide has become an expected (if not accepted) practice. In Cambodia, Pol Pot is said to have killed more people than Adolf Hitler. There, and in Kosovo, and several African countries, pictures of the helpless victims document the crimes.

The religious gap worldwide seems to get ever wider and bloodier. Jerusalem is a tinderbox ready to explode in the strife-ridden Middle East. Hindu and Muslim mobs kill and loot in India. Christian and Muslim gangs engage in bloody fighting in Indonesia. Many victims are stabbed, beaten, or trapped in fires. Others are attacked with rocks, machetes, clubs—even bows and arrows.

In Japan, long the model of order and decorum, the nation's schools are exploding with unrest; 44 percent of elementary and junior high schools report "collapsed classrooms" according to a report issued by the Kyodo News Agency. Once-obedient students fight, walk on desks, spit, urinate on verandas, and talk nonstop during classes. What does this breakdown of civility mean?

Such gaps in world society do not seem to concern many Americans. After all, our stock markets soar, our car sales boom, our rich get richer. We suffer from affluenza—having so much of everything, being so affluent, facing endless glut and credit card debt.

Perhaps more than anything else, it is wealth that divides the world and widens the gap. The United Nations' Human Development Report for 1998 makes this crystal clear. Last year the world consumed more than $24 trillion in goods and services—six times more than the figure for 1975. And things continue to skyrocket in 1999.

But this wealth is not spread equally around the world, or in our own nation. The United States, with less than 5 percent of the world's population, controls a staggering 40 percent of the world's wealth. The three richest people in the world own assets that exceed the combined gross domestic products of the world's forty-eight poorest countries. Hard to believe—but true.

Of the world's 6.8 billion people, over 4.4 billion live today in what we euphemistically call "developing countries." What many are developing is poverty, AIDS, genocide, and despair. The figures on these unfortunate people—numbering in the billions—is both sobering and staggering. They should be contrasted with the "Happy Days Are Here" rhetoric of our politicians—typified by President Clinton's State of the Union speech on January 19, 1999. It's easier to get applause than results if you promise everybody something.

Three-fifths of the world's hapless 4.4 billion have no access to basic sanitation, and a third have no safe drinking water. The world's children suffer the most. One-fifth of them are undernourished and cannot hope to get as far as grade five in school. How will *they* fare in the world of high tech, the World Wide Web, and compulsory computers?

Ah, you say—along with those sharing our Baby Boom Bonanza, many of whom make not thousands but millions of dollars a year—we can't afford worldwide basic education. It would cost $6 billion a year. We might remind them that in the United States alone we spend $8 billion annually on cosmetics. Add these to the list: Europe spends $11 billion annually on ice cream while in Europe and the United States we are spending $12 billion a year on perfumes.

What about the hungry—the myriads that are starving to death with no relief in sight? Again, the plump and prosperous (one estimate has a third of Americans overweight) might point out that basic health care and nutrition would cost a staggering $13 billion. Does it bother them to know that Europe and the United States spend $17 billion per year on pet food?

Even harder to sanction are the billions spent on items that are known to have dire effects on users: $450 billion on cigarettes in Europe (the figure in the United States is not available); $105 billion on alcoholic drinks in Europe; and $400 billion on narcotic drugs worldwide. How do these expenditures affect the gap?

In times of astonishing prosperity and growth in America the utopian dream takes over. Few recall that the word "utopia" comes from two Greek words—*ou* and *topos*—which mean "no place." In the end, Newton was right—all that goes up comes down. The Three Little Pigs also found out that to build our houses on sand and spend our time singing "Who's Afraid of the Big Bad Wolf?" will not keep the wolf away from our door. History happens.

I do not share the Doom and Gloom beliefs of some popular writers; I think the sun will still shine, and Boy Scouts will go on hikes. The problems of democracy are solved not by new restrictions but by more democracy. I will stand fast by the Jeffersonian doctrine: give the people light, and they will find their way. And the people's light comes from and with their popular culture.

Yes, we enter the twenty-first century bewildered and bedazzled. Never has so much changed so quickly for so many. Today's revolutions are more concerned with technique than with ideology. As a metaphor for our time, the computer has no rival. The term "democracy" has been expanded to have not only political but also technological overtones. Yet for all the stunning success of the stock market and the money managers, there is a deep disquiet in our land. A growing number of people think we are raping nature to appease our greed, forgetting the poor so we can get richer and richer. The word "Luddite" is back in style. Our machine culture, theologian Paul Tillich noted, may be less brutal but more effective than totalitarian suppression.

Aldous Huxley is dead, but his *Brave New World* (1932), depicting a land obsessed with sex, sports, and soma, goes marching on. In it, mobs (en route to the stadium? or the mall?) sing their little ditties:

> Orgy-porgy, Ford and fun
> Kiss the girls and make them One.
> Boys at one with girls at peace
> Orgy-porgy gives release.

Massive glut will be inevitable; and clever advertisers will manipulate and exploit. Audiences will be fragmented, quality will be diluted, and technology will be trivialized. Children will be especially vulnerable, as the marketplace becomes their message even before they can read or write. Holed up with a home information center, cut off from community and friends, individuals will be lost in cyberspace, victims of a new pervasive unreality.

Of course we cannot be sure about these things. Revolutionary change is always frightening, especially for those raised in an earlier (print) culture. Machines can't control us if we control machines. We program computers—they need never program us. We don't have to be captive to 200 new channels, many of which ask us to shop and consume. We can turn off the set.

Theories of energy, matter, space, and time are in flux. What was once thought to be simple has become incredibly complex. Our future depends on understanding the past, and those who have shaped and championed it; who have opened their eyes to new truth, but remembered their goodly heritage. They have demanded to be heard. They have been willing to take risks—to take their stand.

I'll Take My Stand is the title of an important book published in 1930, by a group known as Twelve Southerners. They believed in and fought for a Southern regional tradition, heritage, and "way of life" which was under attack. Those "taking their stand" now come from all over the globe. They fear the consumer capitalism that turns our wonderfully diverse planet into a standardized global village. They still fear a crass mass culture that champions conformity while praising diversity. They see the computer as the avant garde of dangerous trends, and refuse to let it dehumanize us. We are not ciphers. We are neither "0s" nor "1s" on silicon chips and will not be computerized. Our world is neither black nor white but full of infinite numbers of grays. We must sail cautiously into treacherous uncharted waters.

If we alone can control our time, not so our privacy. We may want it, but others may want to violate it—and their methods grow ever more insidious, ever more flexible. Once it was a human voice or presence that burst in—like the famous "visitor from Porluck" who interrupted and shorted forever Coleridge's "Kubla Khan." Now the visitor comes via junk mail, unwanted phone calls, phone mail, e-mail, you name it.

But the worst intrusion comes from the new click-and-show computer culture. Our lives are not only open books—they are also on-line. We suddenly realize that much of our lives—including what has always been regarded as "private"—is an open book to those clever enough to peek at our credit ratings, credit card numbers, ATM numbers, tax forms, medical records, and legal transactions. No one seems to know how to stop them. New companies collect "private" information and sell it. Advertisers get valuable information from children, via TV games and the Internet, and then use it against parents. How can this be regulated?

Answering questions such as this might restore what our founding fathers called "domestic tranquillity."

We academics have split into defensive enclaves, politically correct parrots, who live by dissent—about our mission, heroes, texts, methods, and "the canon." The foundation of traditional authority has been deconstructed and hence demolished. All too often all we are left with is a pile of rubble.

Yet from this rubble we can construct a better and brighter world. The message seems clear: we must find both mission and method by which to reach a new consensus and understanding. Construct a new and better world.

We did it before. We can do it again.

Notes

Foreword

1. Mike Wallace, *Mickey Mouse History and Other Essays on American Memory* (Philadelphia, PA: Temple University Press), p. 169.

In the Beginning

1. Robert Hughes, *The Shock of the New* (New York, McGraw-Hill, 1991), p. 13. This book, centering on the hundred-year history of modern art—its rise and dazzling fall—connects elite and popular art, and their interaction, in a uniquely successful way. It has been helpful in my own work and research.

Cavespace

1. Henry Glassie, *The Spirit of Folk Art* (New York, Abrams, 1989), p. 250. This is a pivotal book, and helped establish Glassie as one of the world's leading folklorists. He has since written award-winning studies centering on Ireland, Turkey, and Bangladesh.
2. Ibid., p. 252.

Elvis the Incredible

1. Peter Guralnick's two massive volumes on Elvis are by far the best account: *Last Train to Memphis* (Boston, Little, Brown, 1994) and *Careless Love: The Unmaking of Elvis Presley* (Boston, Little, Brown, 1999). Guralnick's earlier volume, *Free Sweet Soul Music: Rhythm and Blues and the Southern Dream* (New York, Harper & Row, 1986), is also important.

The Return of the Luddites

1. James Combs, "In Praise of Ned Ludd," in *The Global Village: Dead or Alive?* ed. by Ray Browne and Marshall Fishwick (Popular Press, Bowling Green, OH, 1999) p. 211.

2. The matter is fully explored in E. P. Thompson's classic, *The Making of the English Working Class* (New York: Vintage, 1966).

3. Neil Postman, *Technopoly* (New York, Knopf, 1992).

4. James Combs, op. cit., p. 218.

People

1. *Demos* also means deme—one of the hundred townships into which Cleisthenes divided Attica (about 508 B.C.).

2. For more on these problems, see E. J. Hobsbawn, *Primitive Rebels* (Manchester, U.K., University Press, 1959); Richard Samuel, *Village Life and Labour* (London, Unwin, 1975); and M. de Certeau, *L'Ecriture de l'Histoire* (Paris, n. p., 1985).

3. George Dorsey, *The Story of Civilization: Man's Own Show* (New York, Halcyon, 1931), p. 197; and A. C. Spectorsky, *The Book of the Earth* (New York, Macmillan, 1957). See also, W. M. S. Russel, "Plutarch as a Folklorist," in Venetia J. Newhall, ed., *Folklore Studies in the Twentieth Century* (Bury St. Edmunds, U.K., St. Edmundsbury Press, 1980), pp. 371 ff.

4. George Rude, *The Crowd in the French Revolution* (Oxford, U.K., Clarendon Press, 1959), p. 232. See also Rude's *The Crowd in History: A Study of Popular Disturbances in France and England, 1730-1848* (New York, John Wiley & Sons, 1964). Rude builds on the earlier work of Gustave Le Bon, whose study *The Crowd* appeared in 1909. See also Hobsbawn, *Primitive Rebels*; Ernesto Canetti, *Crowds and Power* (New York, Viking, 1962); David Riesman, *The Lonely Crowd* (New Haven, CT, Yale University Press, 1950); and M. J. Smelser, *Theory of Collective Behavior* (New York, Knopf, 1963).

5. David Cort, *Revolution by Cliché* (New York, Funk & Wagnalls, 1970).

6. All these topics are included in Eileen and Stephen Yeo, eds., *Popular Culture and Class Conflict, 1590-1914: Explorations in the History of Labour and Leisure* (Sussex, U.K., Harvester Press, 1981). See also Terry Eagleton, *Marxism and Literary Criticism* (New York, Methuen, 1976). Of many recent Russian studies, Gregori Organov's *Genuine Culture and False Substitutes* (Moscow, Novosti Press, 1979) is perhaps the most comprehensive.

7. Deric Regin, *Culture and the Crowd: A Cultural History of the Proletarian Era* (Philadelphia, Chilton, 1968); see also Raymond Williams, *Culture and Society, 1870-1950* (New York, Harper, 1959); and Hannah Arendt, *The Human Condition* (Chicago, University of Chicago Press, 1958).

8. See Max Horkheimer and Theodor W. Adorno, *Dialectic of Enlightenment* (New York, Basic Books, 1972); and Theodor W. Adorno, *Prisms* (London, Methuen, 1972).

9. The matter is discussed in "The Average Way of Living," in Regin, *Culture and the Crowd*. See also "Mass Culture and Mass Media," in *Daedalus*, Spring 1960.

10. So argues Robert Strausz-Hupe in *The Zone of Indifference* (New York, G. P. Putnam's Sons, 1952), p. 64.

11. That "matrix" was transformed radically by the financial reversals throughout Asia in 1998-1999. The most shocking aspect was the recession in Ja-

pan, the world's second largest economy. This, plus the continuing woes in the former Soviet Union, left earlier Utopian dreams in shambles.

Golden Bangladesh

1. Henry Glassie, *Art and Life in Bangladesh* (Bloomington, IN, Indiana University Press, 1997), p. 231. Glassie's work on Bangladesh is the primary source for students of its history and folklore.

East and West

1. Rudyard Kipling, "The Ballad of East and West," in *Complete Poems.* (New York, Bantam Doubleday Dell, 1994).

Ray and Ronald

1. Marshall Fishwick, *The World of Ronald McDonald* (Bowling Green, OH, Popular Press, 1978). See also George Ritzer, *The McDonaldization of Society: An Investigation into the Changing Character of Contemporary Social Life* (Newbury Park, CA, Pine Forge Press, 1995).

2. The first major study, Max Boas and Steve Chain's *Big Mac: The Unauthorized Story of McDonald's* (New York, Dutton, 1976) is not only unauthorized but bombastic and inaccurate. The independent journalist John F. Love spent four and a half years researching and writing *McDonald's: Behind the Arches* (New York, Bantam Books, 1986). It is much more reliable, if largely laudatory. The *Annual Reports* issued by the McDonald's Corporation are essential for any study. For a wider scope and coverage, see Waverly Root and Richard de Rochemont, *Eating in America: A History* (New York, William Morrow, 1976).

3. Ray Kroc (with Robert Anderson) told his own story in *Grinding It Out: The Making of McDonald's* (Chicago, Henry Regnery, 1977).

4. John F. Love, *McDonald's: Behind the Arches* (New York, Bantam Books, 1986), pp. 1-9.

5. Alan Hess, "The Origins of McDonald's Golden Arches," *Journal of the Society of Architectural Historians*, 45: March, 1968, pp. 60-62.

6. Marshall Fishwick and J. Meredith Neil, *Popular Architecture* (Bowling Green, OH, Popular Press, 1978). See especially Denise Scott Brown's essay on "Learning from Pop," pp. 9-17.

7. Love, *McDonald's,* p. 417 f.

8. Letter to the author dated October 31, 1993.

Note: I am grateful to Teresa Edwards and Christopher Rooney for research and ideas.

9. Ibid.
10. Ritzer, *The McDonaldization of Society,* p. 98.

Fling Wide the Gates!

1. Hugh Thomas, *World History: The Story of Mankind from Prehistory to the Present.* (New York, HarperCollins rev. ed., 1996), p. 708. In this excellent book one of the foremost historians of our time surveys the story of humankind, including interesting material on popular culture. He deals with such matters as food and drink, dress, class and race, cities, and communications.

The Pop Heard Round the World

1. Not all our leading scholars paint a rosy picture. Norman F. Cantor, for example, issues some dire warnings in *The American Century: Varieties in Modern Times* (New York, HarperCollins, 1997). He thinks the optimistic forecast in the 1960s that a new synthetic culture would emerge has not been realized; and that the split between high and popular culture has been a major cause of the intellectual stagnation that set in during the early 1990s (p. 510). Such views contrast sharply with the rosy rhetoric from politicians and business leaders.

Further Reading

With so broad a subject covering not only centuries but millennia, a reading list could become a large volume in its own right. *Books in Print,* for example, contains 1,659 entries under "Popular Culture," with twenty-two subdivisions. Of those no longer in print, there is no end.

Hence, we include here books to supplement key sources already listed in the Endnotes and (on several occasions) Notes sections. Only books and articles of special interest have been selected, and the most recent titles have been given preference.

* * *

Anderson, Walter T. *Reality Isn't What It Used to Be.* New York: HarperCollins, 1990.

Ang, Ien. *Desperately Seeking the Audience.* London: Routledge, 1991.

Bailey, Brian J. *The Luddite Rebellion.* United States: New York University Press, 1998.

Baudrillard, Jean. *The Cool Provocateur.* New York: Verson, 1998.

———. *The Ecstasy of Communication.* New York: Automedia, 1999.

———. *Forget Foucault.* New York: Automedia, 1999.

Baughman, James L. *The Republic of Mass Culture.* Baltimore: Johns Hopkins University Press, 1996.

Bauman, Zygmunt. *Modernity and Ambivalence.* Cambridge, UK: Polity Press, 1991.

Bigsby, C.W.E. *Superculture: American Popular Culture and Europe.* Bowling Green, OH: Popular Press, 1975.

Blackham, H.J. *The Future of Our Past.* New York: Prometheus Books, 1996.

Bloom, Harold. *The American Religion: The Emergence of a Post-Christian Nation.* New York: Simon & Schuster, 1992.

Boas, George. "The People." In *The History of Ideas: An Introduction.* New York: Charles Scribner's, 1996.

Boorstin, Daniel J. *The Image: A Guide to Pseudo-Events in America.* New York: Vintage, 1961.

Browne, Ray B. *Heroes of Popular Culture.* Bowling Green, OH: Bowling Green University Press, 1972.

_______. "Popular Culture as the New Humanities." *Journal of Popular Culture*, vol. 17, no. 4, Spring, 1984.

_______. *Against Academia: The History of the Popular Culture Movement, 1967-1988*. Bowling Green, OH: Popular Press, 1989.

_______. *Dominant Symbols in Popular Culture.* Bowling Green, OH: Bowling Green University Popular Press, 1990.

_______. *Digging into Popular Culture.* Bowling Green, OH: Bowling Green University Popular Press, 1991.

_______. *Continuities in Popular Culture.* Bowling Green, OH: Bowling Green University Popular Press, 1993.

_______. *Pop Culture Landmarks.* New York: Gale Research, Inc., 1994.

Burke, Peter. *Popular Culture in Early Modern Europe.* Harper Torchbooks, 1978. See especially Appendix A: "The Discovery of the People: Select Studies and Anthologies, 1760-1846."

Campbell, Burke. *The Naked Global Village.* New York: Random House, 1997.

Cantor, Norman F. and Werthman, Michael S., eds. *The History of Popular Culture.* New York: Macmillan, 1968.

Carlson, Lew, ed. *American Popular Culture at Home and Abroad.* Milwaukee: New Issues Press, 1995.

Carney, George O. *Fast Food Stock Cars & Rock-n-Roll: Place and Space in American Pop Culture*. London: Rowman & Littlefield, 1995.

Chaucer, Geoffrey. *The Canterbury Tales.* New York: Oxford University Press, 1998.

Chowdhury, Omar. *Prelude to Bangladesh.* New Delhi: South Asian Publishers, 1980.

Collins, Jim. *Uncommon Cultures: Popular Culture and Post-Modernism.* London: Routledge, 1989.

Combs, James and Nimmo, Dan. *The Comedy of Democracy.* New York: Praeger, 1997.

Cook, B.S. *The Elgin Marbles.* London: British Museum Press, 1997.

Crisp, Peter. *The Parthenon: How It Was Built and How It Was Used.* Austin, TX: Raintree, Streik, Vaughan, 1997.

Crowley, David and Mitchell, David. *Communication Theory Today.* Stanford, CA: Stanford University Press, 1994.

Daly, Steven and Wice, Nathaniel. *alt.culture: An a-to-z Guide to the 90's.* New York: Harper Perennial, 1995.

Davis, Robert C. *The War of the Fists: Popular Culture and Public Violence in Late Renaissance Venice.* New York: Oxford University Press, 1994.

December, John, ed. *The World Wide Web Unleashed.* Indianapolis: SAMS, 1995.

Douglas, Ann. *The Feminization of American Culture.* New York: Farrar, Strauss, & Giroux, 1998.

Duffy, Dennis. *Marshall McLuhan*. Toronto: McClelland and Stewart, 1969.

During, Simon, ed. *The Cultural Studies Reader.* New York: Routledge, 1993.

Eco, Umberto. *A Theory of Semiotics*. Bloomington, IN: Indiana University Press, 1976.

Ferguson, Marjorie and Golding, Peter. *Cultural Studies in Question.* London: Sage, 1997.
Firth, Raymond. *Symbols: Public and Private.* Ithaca, NY: Cornell University Press, 1973.
Fishwick, Marshall W. *Common Culture and the Great Tradition: The Case for Renewal.* Westport, CT: Greenwood Press, 1982.
________. *Seven Pillars of Popular Culture.* Westport, CT: Greenwood Press, 1985.
________. *Dominant Symbols in Popular Culture.* Bowling Green, OH: Popular Press, 1990.
________. *Go, and Catch a Falling Star: Rethinking Popular Culture.* New York: American Heritage, 1994.
________. *An American Mosaic: Refiguring American Culture History.* New York: American Heritage, 1996.
________. *Globalism: Dead on Arrival.* Bowling Green, OH: Popular Press, 1999.
Fiske, John. *Media Matters: Everyday Culture and Political Change.* Minneapolis: University of Minnesota Press, 1994.
Gans, Herbert. *Popular Culture and High Culture.* New York: Basic Books, 1974.
Garcis Canclini, N. *Transforming Modernity: Popular Culture in Mexico.* Austin, TX: University of Texas Press, 1995.
Gates, Bill. *The Road Ahead.* New York: Viking Penguin, 1996.
________. *The Secret Diary of Bill Gates.* New York: Andrews & McMeel, 1998.
Giraux, Henry A. *Disturbing Pleasures: Learning Popular Culture.* New York: Routledge, 1994.
Gitlin, Todd, ed. *Watching Television: A Pantheon Guide to Popular Culture.* New York: Pantheon Books, 1986.
Glassie, Henry. *The Old Traditional Way of Life.* New York: Trickster Press, 1989.
________. *The Spirit of Folk Art: The Girard Collection at the Museum of International folk Art.* New York: Harry N. Abrams, 1995.
________. *Turkish Traditional Art Today.* Bloomington, IN: Indiana University Press, 1996.
________. *Art and Life in Bangladesh.* Bloomington, IN: Indiana University Press, 1998.
Goldstein, Jeffrey. *Why We Watch: The Attractions of Violent Entertainment.* New York: Oxford University Press, 1998.
Goodwin, Andrew. *Dancing in the Distraction Factory: Music, Television, and Popular Culture.* Minneapolis: University of Minnesota Press, 1993.
Gordon, Ian. *Comic Strips and Consumer Culture.* Washington, DC: Smithsonian Institute Press, 1998.
Gray, George Z. *The Children's Crusade.* Stark, KY: DeYoung Press, 1997.
Griswold, Wendy. *Cultures and Societies in a Changing World.* Thousand Oaks, CA: Pine Forge, 1994.
Joyce, Patrick. *Visions of the People.* New York: Cambridge University Press, 1994.

Hooper, J.T. and Burland, Cottie A. *Art of Primitive People.* Philadelphia: Philosophical Library, 1954.
Hunter, James D. *Culture Wars: The Struggle to Define America.* New York: Basic Books, 1991.
Inglis, Fred. *Cultural Studies.* Oxford, U.K.: Oxford University Press, 1993.
Jacoby, Russell. *The Last Intellectuals: American Culture in the Age of Academe.* New York: Basic Books, 1987.
Johnson, B.L.C. *Bangladesh.* New York: Barnes and Noble, 1975.
Kaes, Anton. *From Hitler to Heimat: The Return of History to Film.* Cambridge, MA: Harvard University Press, 1989.
Kaufmann, Roland G. *King Tut's Reality: The Seven Steps to Redemption.* Las Vegas: Heridanius Foundation, 1996.
Kittelson, Mary L. *The Soul of Pop Culture.* Chicago: Open Court, 1997.
Kroeber, A.L. and Kluckohn, Clyde. *Culture: A Critical Review of Concepts and Definitions.* New York: Harper & Row, 1963.
Landau, Elaine. *The Curse of Tutankhamen.* Brookfield, CT: Millbrook Press, 1996.
Lapham, Lewis H. *Waiting for the Barbarian.* New York: Verso, 1997.
Lash, Scott and Friedman, James, eds. *Modernity and Identity.* Oxford, U.K.: Oxford University Press, 1992.
Levinson, David and Christensen, Karen. *The Global Companion.* New York: ABC-Clio, 1996.
Marchand, Philip. *Marshall McLuhan: The Medium and the Messenger.* Cambridge, MA: MIT Press, 1998.
McCurdy, Howard E. *Space and the American Imagination.* Washington, DC: Smithsonian Institution, 1967.
McLuhan, Marshall (1911-1982).

Between 1951 and 1995, twenty of McLuhan's books were published, as well as seventeen books about him, plus scores of articles and interviews. They are all listed in Philip Marchand's new biography, (listed above).

Both friend and mentor, McLuhan had more influence on my work than any other person. The most influential three books were: *The Gutenberg Galaxy: The Making of Typographic Man* (1962); *Understanding Media: The Extensions of Man* (1964); and *War and Peace in the Global Village* (1968).

McRobbie, Angela. *Postmodernism and Popular Culture.* New York: Routledge, 1994.
Meyrowitz, Joshua. *No Sense of Place: The Impact of Electronic Media on Social Behavior.* New York: Oxford University Press, 1985.
Mitroff, Ian I. and Bennis, Warren. *The Unreality Industry.* New York: Oxford University Press, 1989.
Monaco, James. *Celebrity: The Media as Image Maker.* New York: Dell, 1978.
Moore, Laurence. *Selling God: American Religion in the Marketplace of Culture.* New York: Oxford University Press, 1994.
Morgan, Hal. *Symbols of America.* New York: Penguin Books, 1987.

Murfin, Ross and Ray, Supryia M. *The Bedford Glossary of Critical and Literary Terms.* Boston: Bedford Books, 1997.

Neuberg, Victor E. Popular *Literature: A History and Guide.* Middlesex, U.K.: Penguin Books, 1977.

Nevitt, Barrington and McLuhan, Maurice, eds. *Who Was Marshall McLuhan?* Toronto: Stoddart, 1995.

Nolan, Riall W. *Building the Global Village.* New York: NAL Dutton, 1999.

Novak, Robert. *King Tut in America.* Fort Wayne, IN: Windless Orchard, 1988.

Nye, Russel. *The Unembarrassed Muse: The Popular Arts in America.* New York: Dial Press, 1970.

Oriard, Michael. *Reading Football: How the Popular Press Created an American Spectacle.* Chapel Hill, NC: University of North Carolina Press, 1998.

Palagia, Olga. *The Pediments of the Parthenon.* Leyden: Brill, 1998.

Phy, Allene Stuart. *The Bible and Popular Culture in America.* Philadelphia: Fortress, 1984.

Polhemus, Ted and Procter, Lynn. *Pop Styles.* London: Vermilion, 1984.

Postman, Neil. *Amusing Ourselves to Death: Public Discourse in the Age of Show Business.* New York: Viking, 1985.

Prose, Francine. *Primitive People.* New York: Ivy Books, 1993.

Real, Michael. *Supermedia.* London: Sage, 1989.

———. *Exploring Media Culture.* Thousand Oaks, CA: Sage, 1996.

Rifkin, Jeremy. *The End of Work.* New York: Putnam, 1995.

Robinson, Frank M. and Davidson, Lawrence. *Pulp Culture: The Art of Fiction Magazines.* Portland, OR: Collectors Press, 1998.

Rollins, Peter C. *Hollywood's World War One.* Bowling Green, OH: Popular Press, 1997.

———. *Hollywood As Historian.* Lexington, KY: University Press of Kentucky, 1998.

———. *Hollywood's Indians.* Lexington, KY: University Press of Kentucky, 1998.

Rosenberg, Harold. "Philosophy in a Pop Key." *New Yorker.* February 27, 1965.

Rosenstone, Robert A. *Visions of the Past: The Challenge of Film to Our Idea of History.* Cambridge, MA: Harvard University Press, 1995.

Roszak, Theodore. *The Cult of Information: The Folklore of Computers and the True Art of Thinking.* New York: Pantheon, 1986.

———. *The Luddite Rebellion.* New York: New York University Press, 1998.

Rowe, David. *Popular Cultures: Rock Music, Sports, and the Politics of Pleasure.* London: Routledge, 1995.

Rubenstein, Ruth P. *Dress Codes: Meanings and Messages in American Culture.* Boulder, CO: Westview Press, 1995.

Rushkoff, Douglas. *Cyberia: Life in the Trenches of Hyperspace.* New York: Ballantine, 1992.

———. *The GenX Reader.* New York: Ballantine, 1994.

———. *Media Virus: Hidden Agendas in Popular Culture.* New York: Ballantine, 1994.

Sanderson, George and Macdonald, Frank. *Marshall McLuhan: The Man and His Message.* Golden, CO: Fulcrum, 1989.

Sandos, James A. and Burgess, Larry E. *The Hunt for Willie Boy: Indian Hating and Popular Culture.* Stillwate, OK: University of Oklahoma Press, 1994.

Savage, William W. Jr. *Commies, Cowboys, and Jungle Queens.* Amherst, MA: University Press of New England, 1988.

Schechter, Harold and Semeiks, J.G. *Patterns in Popular Culture.* New York: Harper & Row, 1980.

Schwartz, Richard A. *Cold War Culture: Media and the Arts, 1947-1990.* New York: Facts on File, 1997.

Scribners Reference Shelf. *Encyclopedia of American Cultural History.* New York: Scribners, 1999.

Shaw, Bud. "The Mission Football," *Cleveland Plain Dealer,* 10 October 1993: 1 D.

Shuman, Michael. *Towards a Global Village.* New York: LPC Inbook, 1999.

Siddiqui, Ashraf. *Folkloric Bangladesh.* Dhaka, Bangladesh: Eden Press, 1977.

Smoodin, Eric, ed. *Disney Discourse: Producing the Magic Kingdom.* New York: Routledge, 1994.

Socolish, R.M. *100 Steps Necessary for Survival in the Global Village.* New York: Seaburn, 1997.

Springhall, John. *Youth, Popular Culture, and Moral Panic: Penny Gaffs and Gangsta Rap, 1830-1996.* New York: St. Martin's, 1998.

Story, John. *An Introductory Guide to Cultural Theory and Popular Culture.* Athens, GA: University of Georgia Press, 1993.

Swiatecka, Jadwiga. *The Idea of the Symbol.* Cambridge, U.K.: Cambridge University Press, 1980.

Taylor, Timothy. *Global Pop.* London: Routledge, 1996.

Tindall, W.Y. *John Bunyan, Mechanick Preacher.* New York, Macmillan, 1934.

Toynbee, Arnold. *Civilization on Trial.* New York: Oxford, 1948.

U.S. Department of State. *Post Report on Bangladesh.* Washington, DC: Government Printing Office, 1981.

Varnedoe, Kirk and Gopnik, Adam. *High and Low: Modern Art, Popular Culture.* New York, Museum of Modern Art, 1990.

Webster, Duncan. *Looka Yonder: The Imaginary American in Populist Culture.* London: Routledge, 1988.

Wellman, Barry. *Networks in the Global Village.* San Francisco: Westview Press, 1995.

Wice, Nathaniel. *Net Chat.* New York: Random House, 1994.

Williams, Raymond. *The Sociology of Culture.* Chicago: University of Chicago, 1995.

________. *The Politics of Modernism.* London: Chapman and Hall, 1996.

________. *Problems in Materialism.* New York, Verso, 1997.

Willis, Susan. *A Primer for Daily Life.* London: Routledge, 1992.

Yoder, Don. *Discovering American Folklife.* New York: Books on Demand, 1997.

Zeldin, Theodore. *An Intimate History of Humanity.* New York: HarperCollins, 1995.
Zinsser, William. *Pop Goes America.* New York: Harper & Row, 1966.

* * *

Newspapers and magazines provide an endless source of material about popular culture, entertainment, and cyberspace. Two central journals are the *Journal of Popular Culture* and the *Journal of American Culture.* Both are published in Bowling Green, Ohio, 43403.

Index